Henry Hodges

Technology in the Ancient World

With drawings by Judith Newcomer

Allen Lane The Penguin Press

First published in 1970
Allen Lane The Penguin Press Vigo Street London W1
SBN 7139 0110 1

Printed in Great Britain by Jarrold and Sons Ltd, Norwich

Contents

Preface

This book is intended to be a straightforward account of the development of mankind's technology from its origins to the end of the period of Roman domination in the opening years of the fifth century A.D. It has been written specifically with the general reader in mind, and for this reason a certain amount of historical and technical paraphernalia has been abandoned. Technological terms have been reduced to the barest minimum, and absolute dates – however approximate – have been used in preference to the period system commonly used in archaeological writing. The normal habit of giving references in the text has also been foregone, since it was felt that little service would have been done by labouring the reader with the mass of periodical literature that this type of reference would have involved. In lieu, a bibliographical note has been added from which the chief sources of reference may be deduced.

It is quite impossible to thank adequately all those who have in one way or another helped me during the preparation of this book, especially the many colleagues with whom over the years I have been able to discuss various problems. To name each and every one would require in itself a volume: to them all I must apologize for this exiguous word of thanks. I must, however, record my indebtedness to Mrs Judith Newcomer for her excellent illustrations; to Miss Sylvia Cookman for the enormous pains she took in acquiring the other illustrations; to Mr J. van de Watering whose labours as designer can only be fully appreciated by those who have seen the disparate parts of the jigsaw with which he had to cope; to Mr James Cochrane, my editor, whose advice and help at every stage in the preparation of this book have been invaluable; to my secretary, Miss Susan Johnson, for preparing the final typescript; and lastly, but in no manner least, to my wife who had the unenviable task of preparing the first draft of the text.

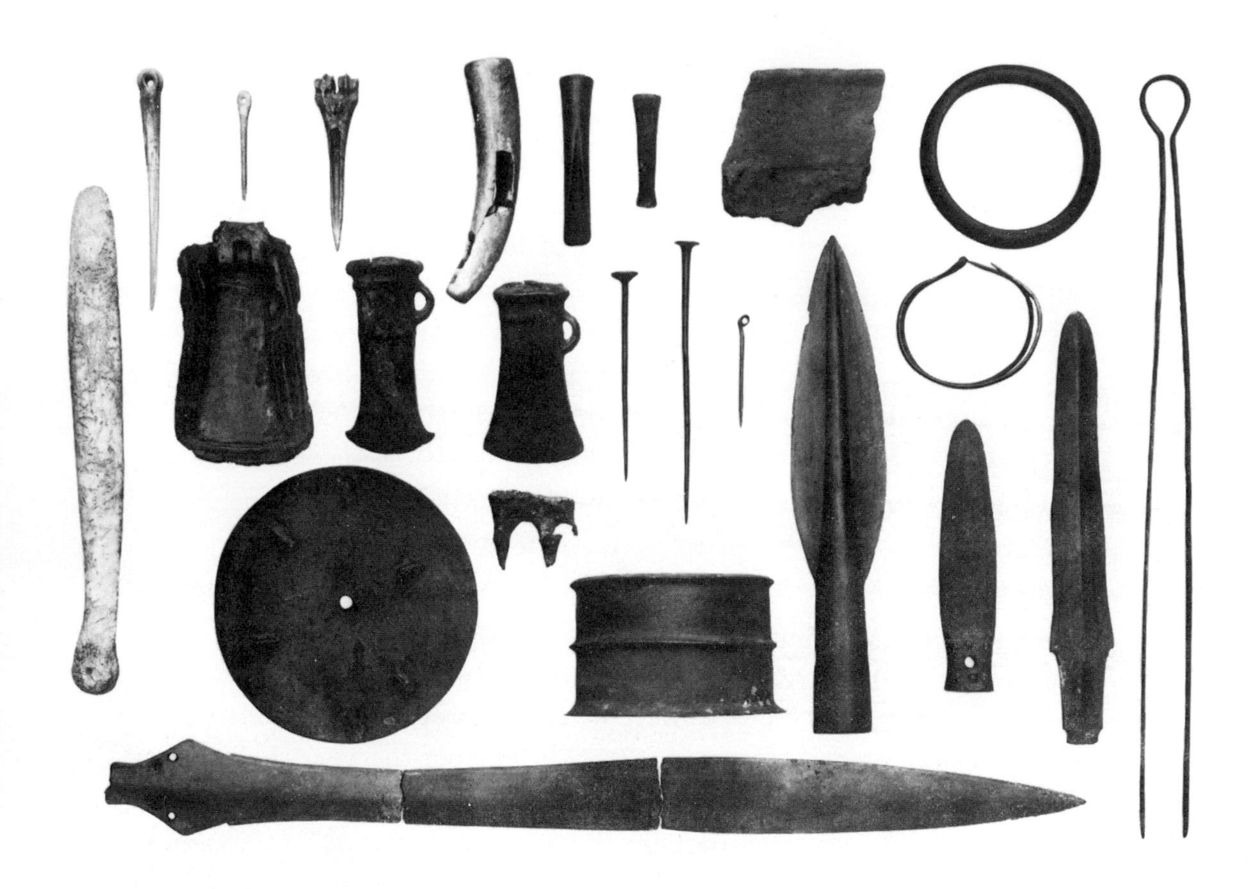

1 Tools and scrap-metal from a cave at Heathery Burn, Durham, about 700 B.C.

Frequently reconstruction of past technologies must depend on fragmentary evidence like the residue from some craftsman's workshop. This group of material, which came from the floor of a cave, contained bronze knives, pins and axes, as well as some bone tools. There was also a sword which had been broken up so that the metal could be used again, and part of a mould (see Figure 61) in which axes were cast. Debris of this sort gives some notion of how the bronze-founder worked, but leaves many questions unanswered.

1
Sources Place and Time

Any account of the early development of technology is bound to resemble a rather poorly made patchwork quilt. The finished fabric will be not only oddly shaped and curiously coloured but also full of holes. For this state of affairs the author is not entirely to blame, for the sources from which he can draw his information are often fragmentary, normally inadequate and frequently misleading. It would be as well, therefore, to begin with an apology and to describe briefly the materials on which the author has had to rely.

The largest single source of information lies in the vast body of excavation reports written by numerous archaeologists over the past century. Sometimes one is fortunate: the excavator, by accident or by design, has tumbled upon the remains of what was once a large site devoted entirely to a single industry, and from his excavation report one gets a very clear picture of, let us say, a pottery works or an iron foundry as it was when being operated in a particular place over a limited period of time. On the whole this is a rare occurrence and more commonly one has to rely on bits and pieces recovered during archaeological excavation. A few corroded tools, a lump or two of the raw materials from which he worked and a handful of rejected scrap may be all that remains from some past craftsman's workshop. Even this is something: far too often one is left with nothing but a blank in the archaeological record, either because the workshops about which one would like to know lie in remote areas where, as yet, little archaeological excavation has been carried out or, as is so often the case, because the equipment used was so perishable that every trace has been lost.

Although archaeological excavation has revealed fewer workshops and even fewer factories than one could wish, nevertheless nearly every object unearthed from the past can be made to tell us something about its method of manufacture, and indeed it is only because of the innumerable painstaking examinations of an enormous range of objects

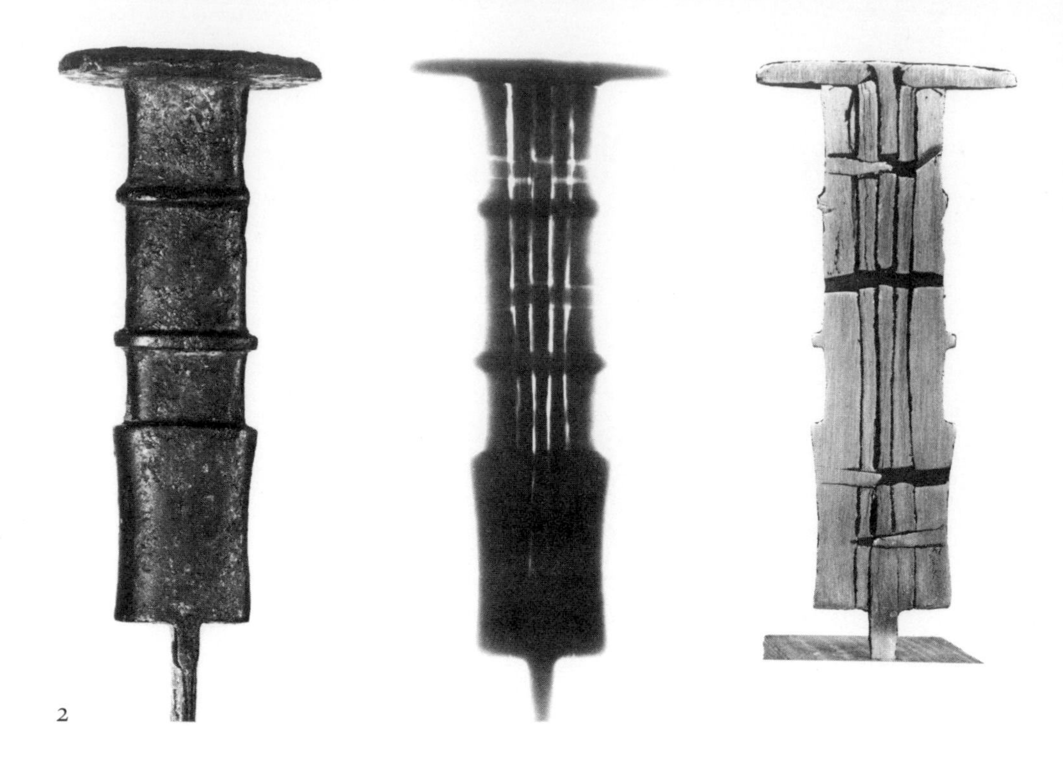
2

made by archaeologists and by scientists that one can even begin to write a history of early technology. The information to be obtained from this source is extremely varied. On the one hand we have hundreds, if not thousands, of chemical analyses of materials such as glass and metals which demonstrate in general terms what early man was using in the manufacture of tools and other equipment. On the other hand are reports, often running into many pages, of the examination of single objects many of which show amazing complexities. Even here, however, the record is strongly biased. This is not a matter of dishonesty amongst archaeologists and scientists. It is the result of the durability and lack of durability of certain classes of material. Pottery and glass, for example, will often survive under the most extreme climatic conditions, while textiles, wood and leather are extremely prone to decay. It follows that one knows far less about the perishable materials than one does about those which are not. Equally, there are countries, of which Egypt is undoubtedly the most important single example in the Old World, where conditions of burial allow even the perishable materials to survive in fairly large quantities. There is a temptation, therefore, to stretch the information to fill the gap, to suppose that what held good in ancient Egypt also held good in ancient Mesopotamia, however often we remind ourselves that the environment, the raw materials, the very tradition of craftsmanship in the two areas were

2 The grip of an iron sword from Luristan: as it appeared to the naked eye; as seen by X-ray; and as it appeared when cut for examination by a metallurgist. Persia, about 800 B.C.

Many objects from antiquity can, if given careful scientific examination, be made to reveal aspects of the techniques used in their making. This iron sword-grip appeared to be made of a single piece of metal, but X-ray photography showed it to be made of many sections held in place by rivets. Later it was cut in half so that an analysis could be made of the various pieces of metal.

3 A group of carpenters' tools from Egypt, about 1500 B.C.

Because of the extreme dryness of most Egyptian tombs, objects made of wood and metal have often survived there, while elsewhere the wood might have rotted away and the metal have become corroded beyond all recognition. As it is, these Egyptian carpenters' tools with copper blades and wooden handles give us a very clear picture of how the craftsmen worked. How each tool was used can be seen in the wall painting (Figure 98).

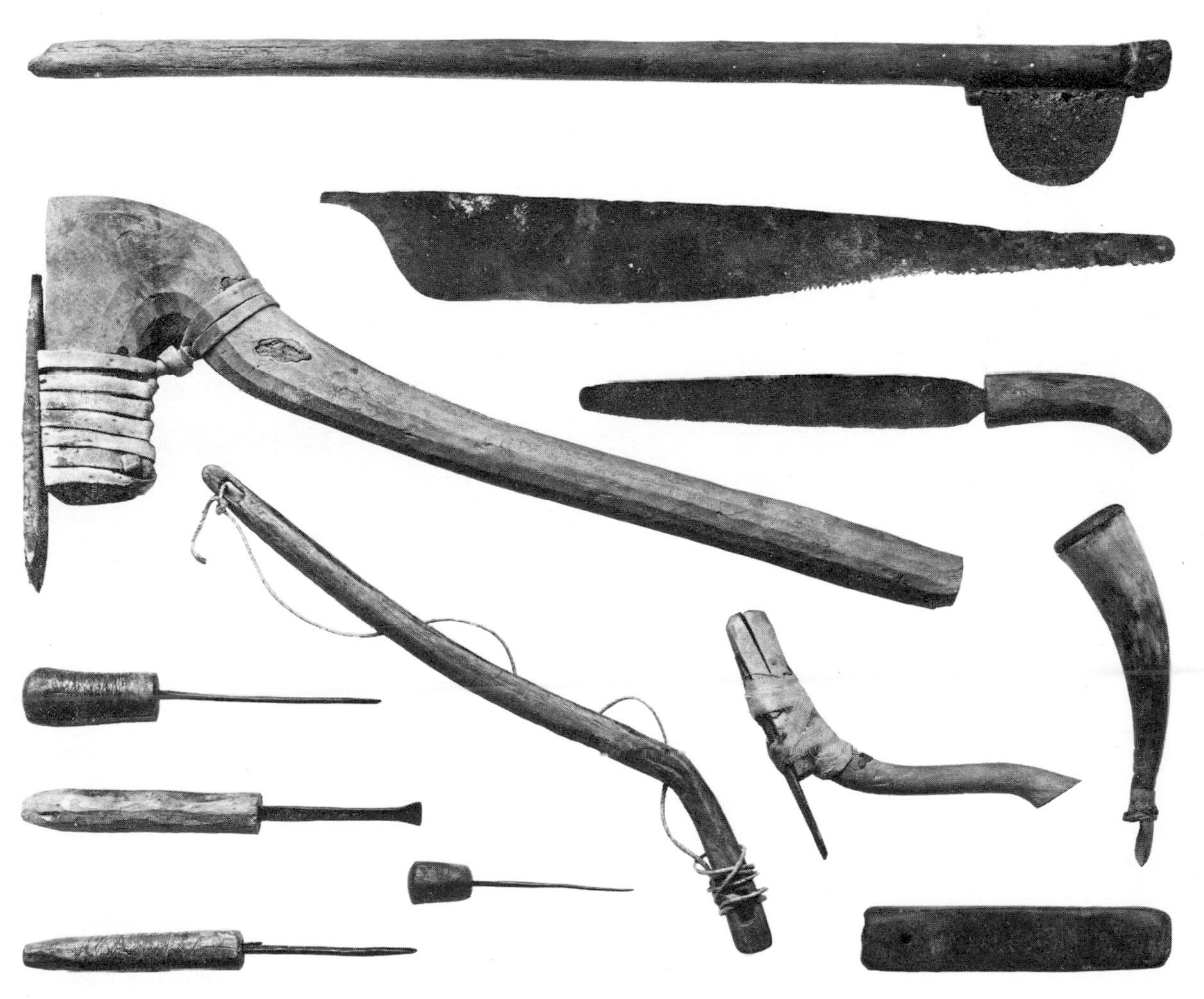

3

always different. It is indeed very frustrating to have to say that one knows nearly nothing about how the early Mesopotamian cobbler set about making a pair of shoes, while one has a great deal of information about the way his Egyptian counterpart worked. It is thus inviting to think of the whole of the Near East as peopled with cobblers working along Egyptian lines instead of leaving gaping lacunae in our history, as in fact one should.

A third source of information lies in the pictures created by artists in antiquity. These may be of widely differing kinds such as wall paintings made in tombs, low reliefs cut into the walls of temples and palaces, mosaic pavements or pictures painted on pottery. Many of these illustrations are highly informative, and many, alas, are highly infuriating. When an Attic vase painter has illustrated a potter at work with a sketch on the side of a jug, one can be pretty sure that his drawing is

4 Transporting a colossus: part of the decoration from the walls of a tomb, Egypt, about 1800 B.C.

A great deal of information about technological processes comes from models and illustrations – paintings, reliefs, mosaics and sculptures – made in antiquity. This illustration from an Egyptian tomb shows, for example, exactly how a colossal statue was transported: the colossus roped to the sledge; the teams of men pulling it; one man pouring water to lubricate the runners; others bringing further supplies on shoulder-yokes, and even the foreman using clappers to give a rhythm to those doing the pulling.

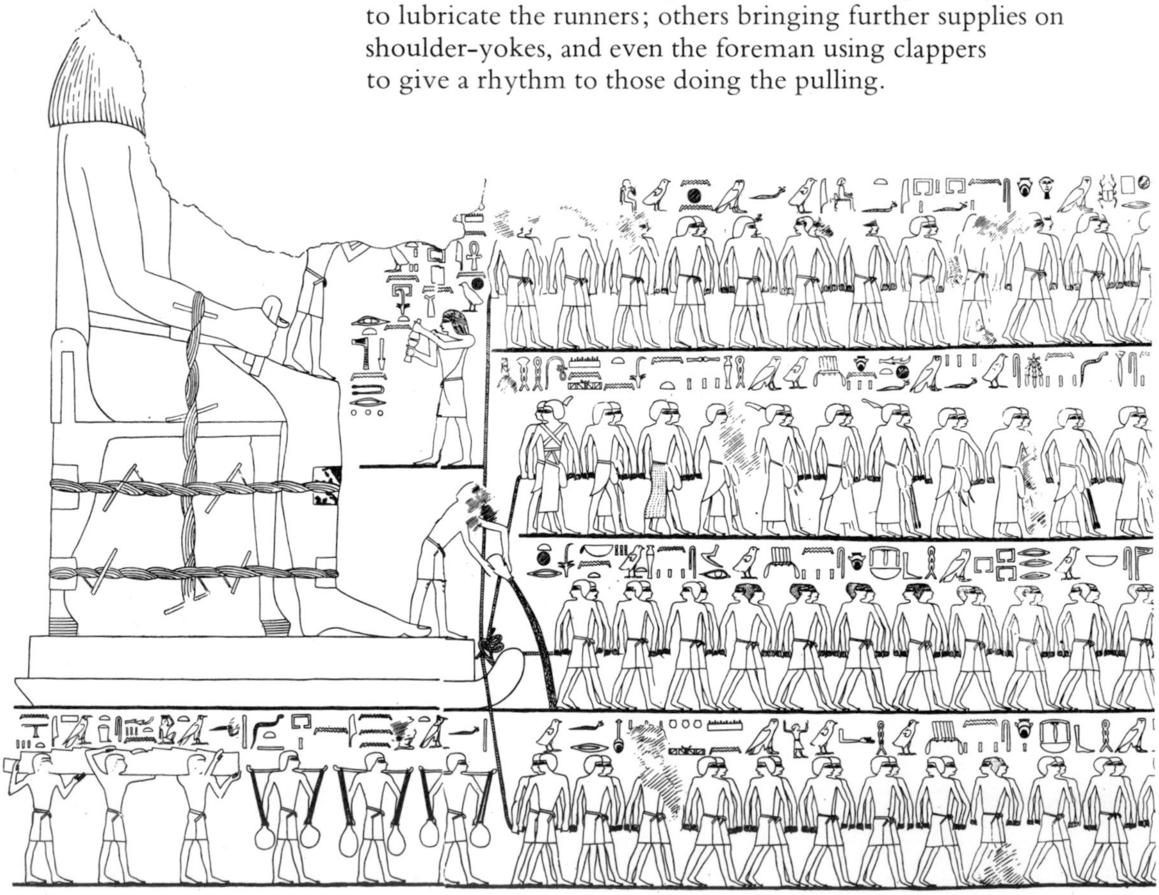

5 Terra-cotta model of a plough-team from Cyprus, about 2200 B.C.

Not all illustrative material is as informative as the foregoing picture from Egypt. This clay model from Cyprus, for example, while clearly showing an ox-drawn plough, gives one little idea of how the plough was constructed or the exact shape of the yoke. Nevertheless, it does tell us that a plough with a single handle was in use on the island about the year 2000 B.C.

right in every practical detail, but when the same artist paints a ship, for example, one is often left in considerable doubt about the accuracy of his detail. Indeed it is only too painfully clear that some vase painters had no idea either how a ship was constructed or how it was sailed. What they have given us is their impression of a ship at sea and not a blueprint from a naval architect. This fact itself is an interesting comment on the diversification of craftsmanship within that period, but it hardly helps us reconstruct early Greek shipping. Unfortunately, from our point of view, many artists were prepared only to give their impressions rather than accurate detailed drawings of things seen, and many, too, had to distort what they were drawing to fit some specific area of the surface they were decorating. Furthermore, one gets the impression that many artists were not free agents and that they had to illustrate what they were told to. As a result, for example, we have plenty of pictures of Mesopotamian monarchs hunting lions and conducting military campaigns, but for the humbler crafts there is an extraordinary blank. By contrast, one feels at times that the Egyptian rulers and aristocracy went out of their way to ensure that the people of later ages knew precisely how every craft was carried out, and that every worthwhile

6 An enigmatic scene: the impression made by a stone seal from Mesopotamia, before 2000 B.C.

Some pictures may be misleading. This seal impression has been variously described by different scholars as showing people preparing food, a banquet in progress, or women making pottery. Obviously all three interpretations are possible and it would be unwise to use the scene as evidence of any one.

7 A Greek drinking-cup depicting two ships at sea, about 600 B.C.

A further difficulty of interpreting pictures from antiquity lies in the fact that artists often distorted what they were illustrating. This was frequently due to the shape of the area that had to be decorated. In this example the hulls of the two ships have had to be curved to fit the surface of the bowl, while the yard-arms have been drawn so as to follow the line of its rim.

technology was depicted stage by stage. Because of this the historian, too, may become unwillingly and unwittingly biased, for it is difficult not to give greater emphasis to the illustrative matter from a country where the paintings and low reliefs are so informative than to a country where apparently the artists were neither allowed nor encouraged to illustrate the crafts.

Finally there are the written sources, and these are almost more variable than any so far considered. At the one extreme lie the very early clay tablets on which the scribes kept the accounts of their masters, and, at the other, we have handed down to us the writings of the Greek and Roman men of science. Clay tablets were written simply as a means of keeping an account of what was what. Thus one finds records of, for example, the number of slaves employed by such and such a person, or the record of the number of chariots stored in the royal palace. There is no reason, therefore, why the early clay tablet records should be of any interest to us at all, and certainly no one could say they were a storehouse of information. But the fact remains that in the whole host of detail one can occasionally pick out rare pieces of information that help us to understand the technology of the period. Thus, for example, one of the scribes receives fifty measures of gold. He records this fact and sends the gold to the foundry for purification. In due course the gold is returned and now weighs only forty-seven measures. Like a good civil servant, to keep his name in the clear he records this fact. Of course one is not told what process was involved, but at least one knows that some sort of refining was being carried out, and at the same time that the process must have been sufficiently well understood for the scribe to be able to persuade his master that everything was in order. Clearly, by the time this record was made the refining of gold was fairly well advanced as a technique.

As the ability to read and write became more widespread, so we find more and more records that are pertinent to our subject, until there emerge at last what can best be called workshop recipes, often quite detailed, giving, for example, the formulation of a particular glass, stating not only the materials to be used but also the processes to be followed. Unfortunately even when one has these records one is not always in the clear. No doubt, to the writer, an item such as 'green stone from the Island of Birds' was perfectly intelligible as an ingredient in glass-making, but we, alas, can identify neither the Island of Birds nor the green stone to be found on it. One fares better when one comes to the writers of classical antiquity. Herodotus, the Greek historian and geographer, writing in the fifth century B.C., provides us with a host of interesting detail although always, of course, as an aside to his main theme. But a lot of what he records was clearly never seen at first hand, and he is apt to prefix a statement with 'they say' and from then on to recount some fantastic fairy-tale. One is at pains to know how much of what Herodotus recorded was information that he acquired himself, how much was just gossip and how much travellers' tales. Even Pliny the Elder, writing in the first century B.C., is not totally reliable. His *Geography* is a remarkable compilation, but if one were to follow some of the instructions he gives for technical processes one would finish up with nothing but a horrid mess. One is left with the impression that, while he recorded accurately as well as he could, he never really followed any technological process through in detail or clearly understood it himself. Nevertheless, the situation is greatly simplified since the writings of classical antiquity have received a great deal of study from innumerable scholars, and it is safe to say that we understand the main content of most of what they have written and what reliance can be placed upon it. The debatable passages in their accounts and their own misunderstandings can, therefore, largely be discounted.

It follows that, no matter from which of the major sources our information comes, our history will be mostly a matter of reconstruction. The materials unearthed by archaeologists, the pictures and the writings, all require interpretation. It is thus extremely difficult to write a purely objective account of the development of early technology, and the author is making no such claim. The reconstructions in this book are the author's own personal view of what things may have looked like or how processes may have been carried out, and new discoveries in the archaeological field any day could force him to change his view completely about many topics here discussed.

Before discussing the development of early technology, however, something further must be said about the background against which this development took place, for we have to reconstruct not only the processes involved in the development of early technology but also the environment in which man was living at that time if we are to understand fully how each step followed the last. Most of the major technological innovations of antiquity were made within the limited area of the Near East and the eastern end of the Mediterranean, and little could be more fatal than imagining that those regions were in antiquity as we know them today. Even in the past ten thousand years enormous changes have taken place which owe nothing to population changes (either migrations or explosions), nor to the recent development of cities, roads and railways. Far more fundamental is the fact that the entire ecology of the region has undergone drastic changes. What we know today as open, dusty plains or rich farmlands were, ten thousand years ago, more or less thickly forested, and within the forest lived a wide variety of wild animals. This is not to say that deserts did not exist, but rather that many hills that we know of today as barren ranges of rock were then at least lightly covered with trees, while the river valleys probably carried very dense forest cover. How the changes from this type of environment to that which we see today took place is a matter of considerable debate amongst scholars. Many see these alterations as a matter of climatic change resulting in a general desiccation of the area; a lowering of rainfall with the subsequent diminution of forest cover. Others see them as a result of man's activities, principally the felling of trees for fuel and building, making such inroads into the forest that eventually an insufficient number of trees were left for regeneration. Others again blame sheep and goats, which in large flocks nibbled away at the young shoots of trees and seedlings, so preventing the regrowth of forest and causing its ultimate destruction. Clearly all three were contributory causes. While in different areas of the Near East it is probable that these three factors made their impact to varying degrees, the sum effect was identical.

If the presence of a reasonable level of tree cover was a prerequisite for the development of, for its day, an advanced technology, it was not a cause, for there are many well-wooded areas of the world where technological advance was, and still is, incredibly slow. One is tempted to ask, therefore, why the Near East was so favoured. Three features of the region stand out above all others – the climate, which offers no real extremes, the morphology, which under primitive conditions allowed fairly easy transport and communications, and, probably most important of all, an enormous diversity of natural resources.

To deal with the last point first, when travelling in the Near East today, despite the arid conditions, one cannot fail to be impressed by the ever-changing pattern of wide river valleys, sweeping highland plateaux, low ranges of rolling hills and towering ranges of mountains. In antiquity each would have carried its own particular flora, and the rapidly changing pattern of light and heavy forest, interspersed with savannah, meant that to many people there was available an enormous range of plant materials, as well, of course, as the animals that dwelt in each different environment. But this is only part of the story. The underlying rock, the cause as it were of the landscape, is equally varied, giving rise to a wide range of possible mineral resources. Thus in many

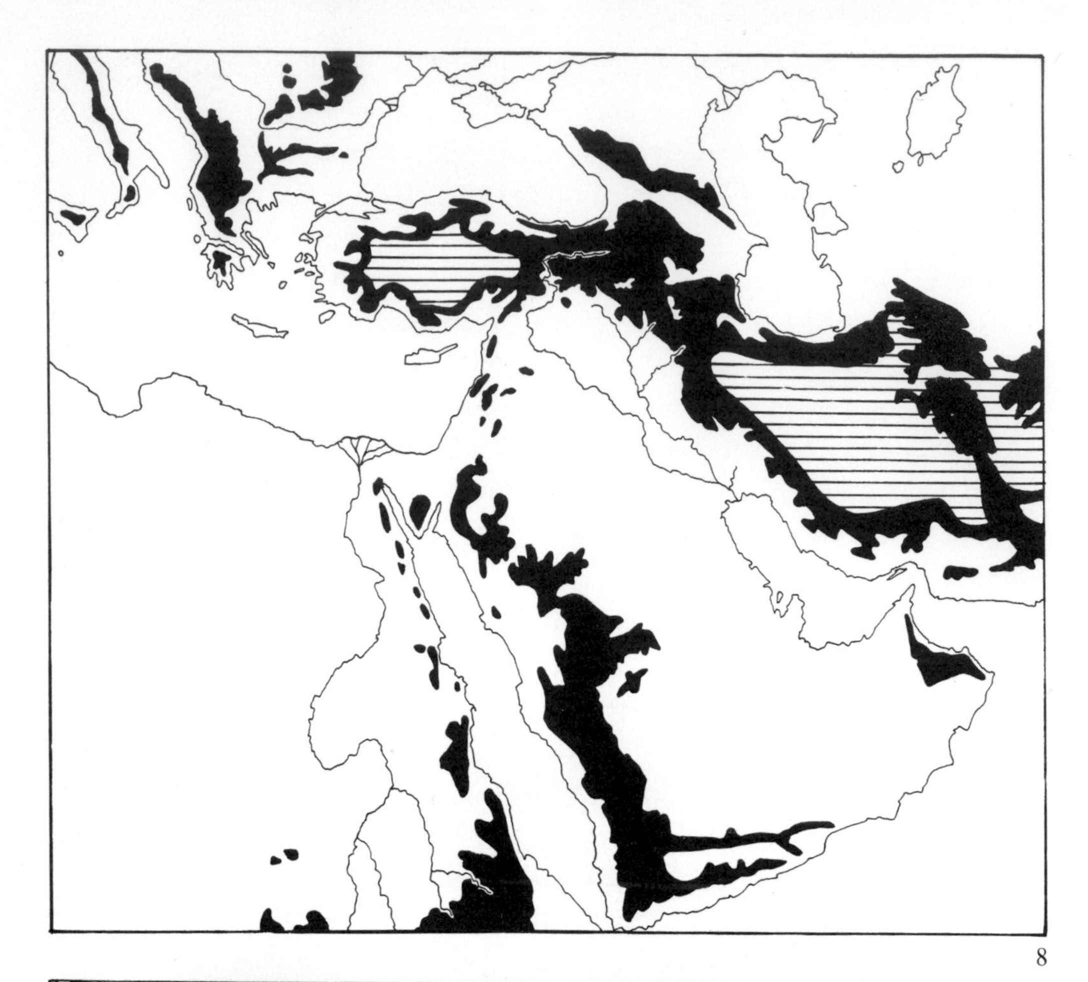

8

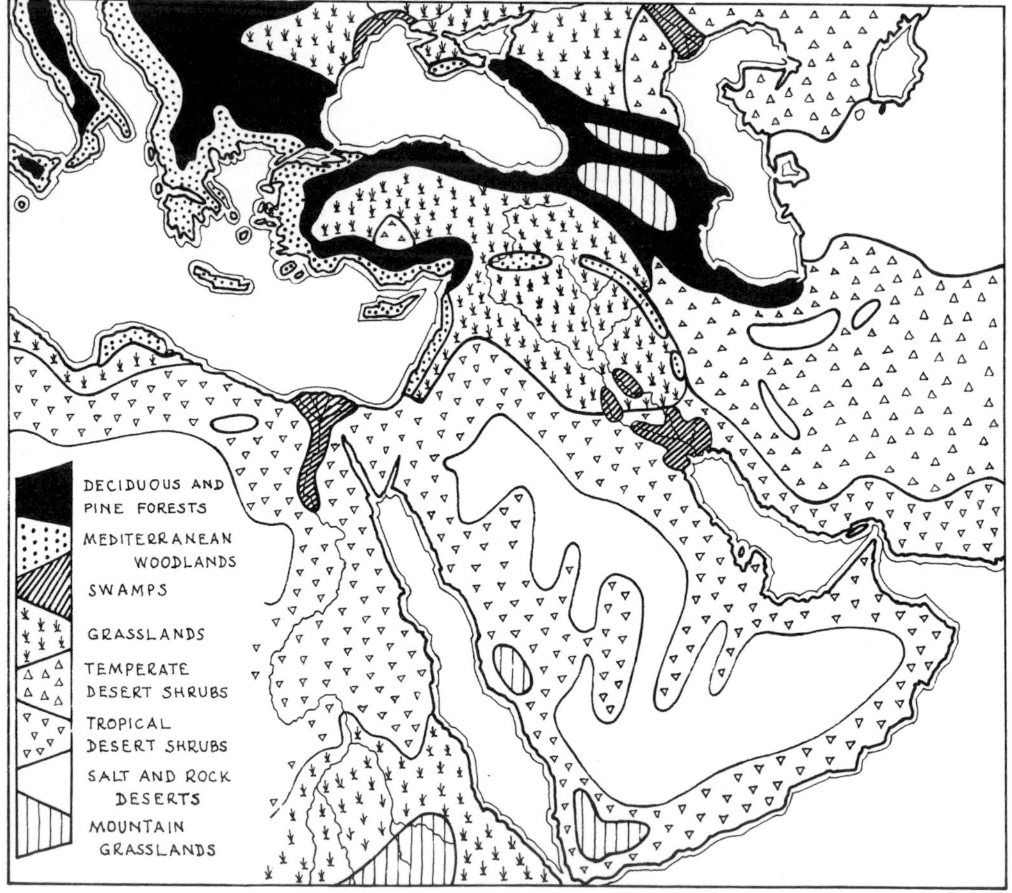
DECIDUOUS AND PINE FORESTS
MEDITERRANEAN WOODLANDS
SWAMPS
GRASSLANDS
TEMPERATE DESERT SHRUBS
TROPICAL DESERT SHRUBS
SALT AND ROCK DESERTS
MOUNTAIN GRASSLANDS

9

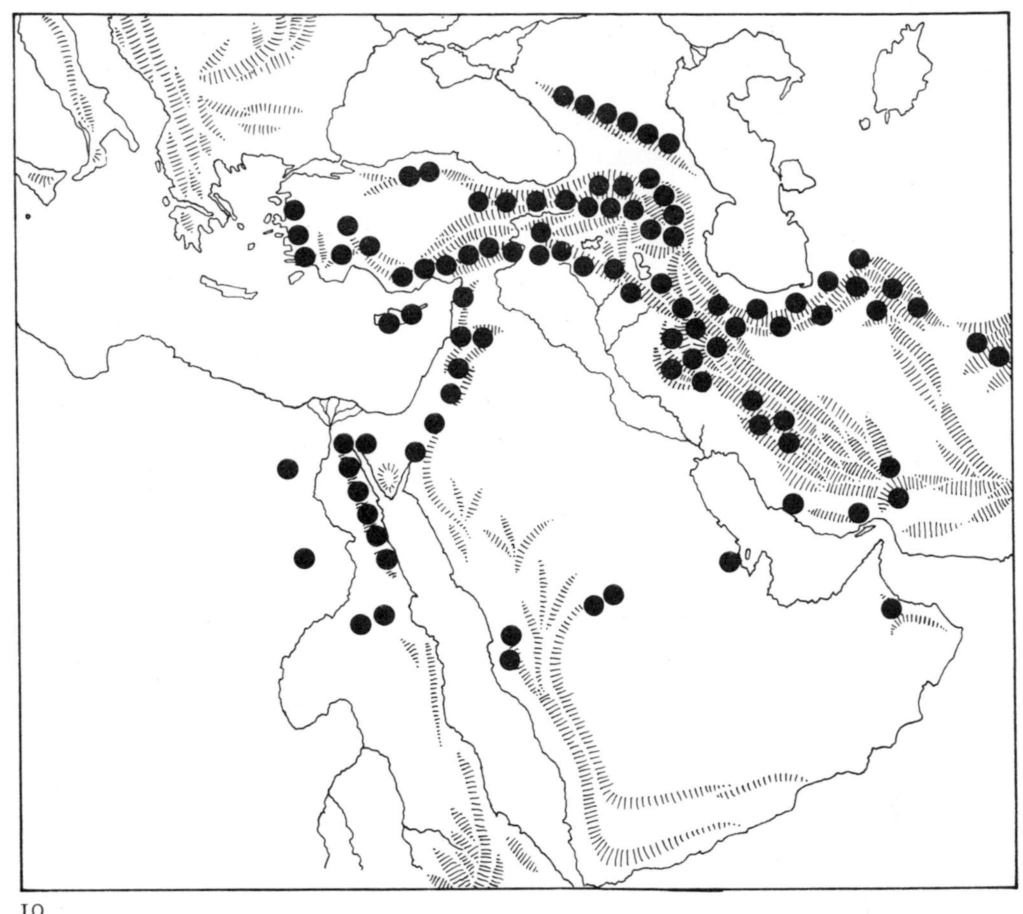

10

8 Topography of the Near East

Notice particularly the two high plateaux of Anatolia and Iran with their surrounding mountains, and the two great river systems, the Euphrates and the Nile, which draw their headwaters from temperate and tropical mountainous areas respectively.

9 Natural vegetation of the Near East

Notice that there is far more extensive woodland than is present today, and also that some regions, as for example Syria, were able to draw upon two or even three quite different vegetational environments.

10 The mineral resources of the Near East

Notice the association between minerals and mountainous areas. The Nile and Euphrates Valleys have been called the 'Fertile Crescent'; their adjacent mountains can with justification be called the 'Mineral Crescent'.

of the hills and mountain ranges are to be found some, if not all, of the ores of the common metals, while the softer and therefore more easily carved stones so suitable for building, such as gypsum and calcite, are even more abundant. There is, too, a more limited number of rare mineral deposits to be found in specific areas, as for example soda and potash in various forms. A wealth of biological and mineral resources, however, is insufficient to account for the early development of technologies in this area, for without adequate communications probably little advance would ever have been made. Small isolated communities tend to tap only a fraction of the available resources, since a large part of technological advance depends upon the interchange of ideas, of the borrowing of a technique or material from one technology and applying it to another.

The rivers and the sea provided both an easy means of transport and a stimulus to the development of boats and shipping, while before the development of water transport the river valleys themselves would have provided early man with a highway. Of the rivers of the Near East two are particularly important to us, the Euphrates and the Nile. From its source in what is today the eastern boundary of Turkey, the Euphrates runs parallel to the shore of the Black Sea for a considerable distance, within 70 miles of the coast, and then turning south passes within 120 miles of Antioch, finally to turn south-east and meander down to the broad alluvial plain of Mesopotamia. The one river, thus, all but links the Black Sea, the Mediterranean and the Persian Gulf, and serves as a link between three very important areas – the mountainous region of Eastern Turkey, rich in mineral resources, the coastal belt of the Eastern Mediterranean and the fertile valley of the Tigris and Euphrates. This does not mean that the Euphrates was navigable as a river throughout its entire length. Indeed, once out of the alluvial plain, the river is full of rapids and can only be used locally for river traffic. Nevertheless the river valley could have served as a pathway throughout the mountainous upper reaches, and the same is true of the Tigris and its tributaries in the Zagros Mountains and in the mountainous area of Kurdistan. Finally, the delta of these two rivers ten thousand years ago was not as we know it today, for a mass of alluvium brought down by the rivers has been deposited at the delta which has thus grown into the Persian Gulf, so that many ancient cities that once stood on the coast, or on the estuary of these rivers not far from the coast, are now many miles inland.

The Nile, by contrast, presents a very different picture, for its headwaters lie in equatorial Africa, climatically a very different zone from its lower reaches. Thus, although we know today that the sources of the river lay near to mountainous areas rich in mineral resources these were apparently never tapped in antiquity; and to the people living in the lower reaches of the Nile, the river, instead of serving as a pathway to an area of wealth, led to a hot and sticky nowhere – a region into which the Egyptians ultimately led a number of expeditions, but not one with which trade was carried out as a normal, everyday affair. For those living in the lower reaches of the river in antiquity, neither the Sahara to the west nor the rocky Arabian Desert to the east, even if then more endowed with plant life than they are today, can have been exactly inviting, least of all suitable for agricultural pursuits. Nevertheless, both the Arabian Desert and Sinai provided the people of the Nile Valley with a wide range of mineral resources, and they thus seldom had to search far afield for the materials they required, while boats and rafts on the river provided an ideal system of transport at least below and between the cataracts.

Other countries of the Eastern Mediterranean, Greece and the Greek islands, the Mediterranean coast of Anatolia and the Levant, present a rather different picture. The mountainous landscape of the Greek mainland, largely of porous limestone, allows no major river system, and what rivers there are provide only small isolated patches of alluvium on which it is possible to grow crops. Contact by overland routes was difficult. Thus we need hardly be surprised that shipping became the most important means of transport and communication. The Mediterranean climate, however, did allow the cultivation of several important crops such as the vine and the olive. These, the abundance of timber, and excellent stone for building may be said to be the only major resources of the area. Cyprus and many of the Greek islands, on the other hand, belong to a different geological formation and were endowed with a fairly wide range of minerals including, most important of all, the ores of copper, while the Anatolian coast and the Levant were doubly blessed, for not only had they a Mediterranean climate and easy access to the sea but also they could draw on the wealth of their hinterlands.

So far we have spoken of ten thousand years ago almost as though it were a starting-point for our history of technology, and in a manner of speaking this is true. But something must now be said about the time-scale against which this history of technology must be projected. At a date that may be reckoned at roughly 10000 B.C. the last of a long series of world-wide major climatic changes had taken place. The great ice-sheets that before this had covered a large part of the Northern

Hemisphere had withdrawn to a region approximating very crudely to that which they occupy today, and accompanying this withdrawal of the ice-sheets were major changes in the flora. Areas peripheral to the ice-sheet that had been tundra now became forested with pine. Areas to the south of this that had been covered with pine now slowly changed to deciduous forest, while tropical forests gradually spread to eliminate some of the southern areas that had once been deciduous. Until these changes had taken place or were well under way, mankind appears to have been only a hunter living by collecting fruit and berries, by fishing and by the taking of game. By about 8000 B.C. these major floral changes were largely complete – at least in the area that we are considering – and at about this point in time we begin to detect the emergence of a new mode of living. Mankind began to explore his environment in a different way, so that ultimately there was to emerge a way of life based upon the domestication of plants and animals, and as may be imagined this transition was accompanied by a considerable number of technological innovations. By the middle of the fourth millennium B.C. – that is to say by about 3500 B.C. – we find mankind taking another step which was ultimately to lead to the development of the City. We can for the moment beg the question as to what is meant by 'City', and pause only to point out that early cities were not just overgrown villages. We shall see later that they were something far more complex, and that the very complexity was engendered by and gave rise to further technologies, of which the winning of copper from its ores and the making of copper alloys was one. And because the alloy of tin and copper, bronze, became the universally accepted metal from which to make tools and weapons, this period is generally referred to by archaeologists as the Bronze Age.

For approximately two thousand years the cities of the Near East grew and became more complex so that every major city increasingly extended its area of influence over an even larger region. By the middle of the second millennium B.C. – that is to say, around the year 1500 B.C. – we find that the cities had grown so large and prosperous that their rulers were able to carve out what could best be termed 'little empires'. The introduction, about the year 1000 B.C., of iron as a metal from which to make tools and weapons added some impetus to this development, but in the first few centuries of the last millennium B.C. shipping had become so advanced as a means of transport that those countries of the Eastern Mediterranean with a coastline found themselves naturally favoured, for they were able to import nearly all those materials they required for their technologies. By 700 B.C. the expansion of trade had been such that we find countries minting their own coins as a means of exchange.

The shift of power to the centre of the Mediterranean, specifically to Rome, can be seen as an extension of this same process, for Barbarian Europe to the North, if having little to offer in the way of advanced technologies, had an enormous reserve of raw materials which could be acquired at a lower cost than from the countries of the Near East. After five centuries of Roman domination, however, the Empire in Europe was overrun by Barbarians, but few if any major technologies were lost to the ken of man, and the world did not thereafter fall into a technological doldrum. Byzantium, Islam and the empires of medieval Europe were all to make their own contributions to a pool of technical skills that were to lay the foundations of the technologies we know today. But the ancient world under the domination of Rome had in fact reached a kind of climax in the technological field. By the end of the Roman period many technologies had advanced as far as possible with the equipment then available, and for further progress to be made a bigger or more complex plant was required. Despite the fact that the Romans were quite capable of indulging in gigantic undertakings, their technologies remained at the small-equipment level. Thus, for example, if it was required to increase the output of iron the number of furnaces was multiplied, but the furnaces themselves remained the same size. Whatever the cause, the idea of building a larger furnace and devising machinery to work it seems to have been beyond the Roman mind. As a result, the last few centuries of Roman domination produced very little that was technologically new. No new raw materials were discovered, no new processes invented, and one can indeed say that long before Rome fell all technological innovation had ceased.

It would be wrong to think that even within the limited area of the Near East and the Eastern Mediterranean technological development was anything like even. A new technology might bear roots in one limited region, and then be brought to something like fruition before it spread to other areas, and even then by some people it would be adopted immediately, while by others it would take an incredibly long time before it was accepted. Sometimes this delay was demonstrably due to a lack of resources in the country adopting the technology, but often it was due to historical events about which, alas, we know very little. Thus, for every technological development there are two important phases and two points in time that we would like to record. First, there is the place and time at which and in which the technology was first developed; and secondly, there is the point in time at which the same technology reached other areas beyond its cradle. But this is an historian's ideal which, for the early periods at least, is seldom to be achieved. The lack of precision in our methods of dating and our patchy knowledge of the past often make it impossible to say precisely where any particular development first took place, so much so that our knowledge that the technology existed at all frequently begins only when it has entered the second phase. A great deal of time can be wasted in speculating about precise origins, particularly where the evidence is meagre. The results are often less credible than the deliberate fictions in the *Just So Stories* or the *Tales of Kai Lung*. If, therefore, new materials and new technologies suddenly appear out of the blue in the following chapters the reader must take it for granted not that this was how they developed but that in fact we know so little of their early origins that it is futile to attempt to record them here.

2
Genesis

(?-5000 BC)

One suspects that, were we confronted by a collection of the earliest tools ever made by man, we would in all probability fail to recognize them for what they were. Such a hypothetical collection might contain a stick chewed to a rough point to be used for digging in search of roots or grubs; a stone roughly chipped along one side to give a crude cutting edge, a tool improvised on the spur of the moment in order to cut up the carcase of some dead beast; or a club made from a branch torn from a tree. We can only speculate that early man made such tools because we know that on occasion apes do so. Implements of this kind made in response to an immediate situation are usually discarded after they have been used, and they thus lack the only feature that would allow us to decide at this point in time that they were in fact tools – they lack any traditional form. Today we can only recognize the earliest tools of man when they unquestionably show a continued tradition in their making. Thus, although by some recent estimates man may have emerged as a distinct species some twenty million years ago, we can only carry our history of tool-making back two million years, while even then our knowledge is limited to one particular type of implement made of a stone that would fracture to give sharp cutting edges. Time has robbed us of the possibility of knowing whether at this remote period mankind was also making wooden tools along traditional lines.

Starting with a lump or pebble of flint, or any other suitably fine-grained rock that would give a sharp edge, a series of flakes was detached by hammering with a second stone in such a way that there emerged an implement with a jagged cutting edge down each side, a crude point and a rounded, smooth butt. The function or functions of this tool, the so-called hand-axe, are unknown – although in all probability it served more or less as does a pocket-knife today, as a tool for general and multiple uses. During the thousands of years in which this tool

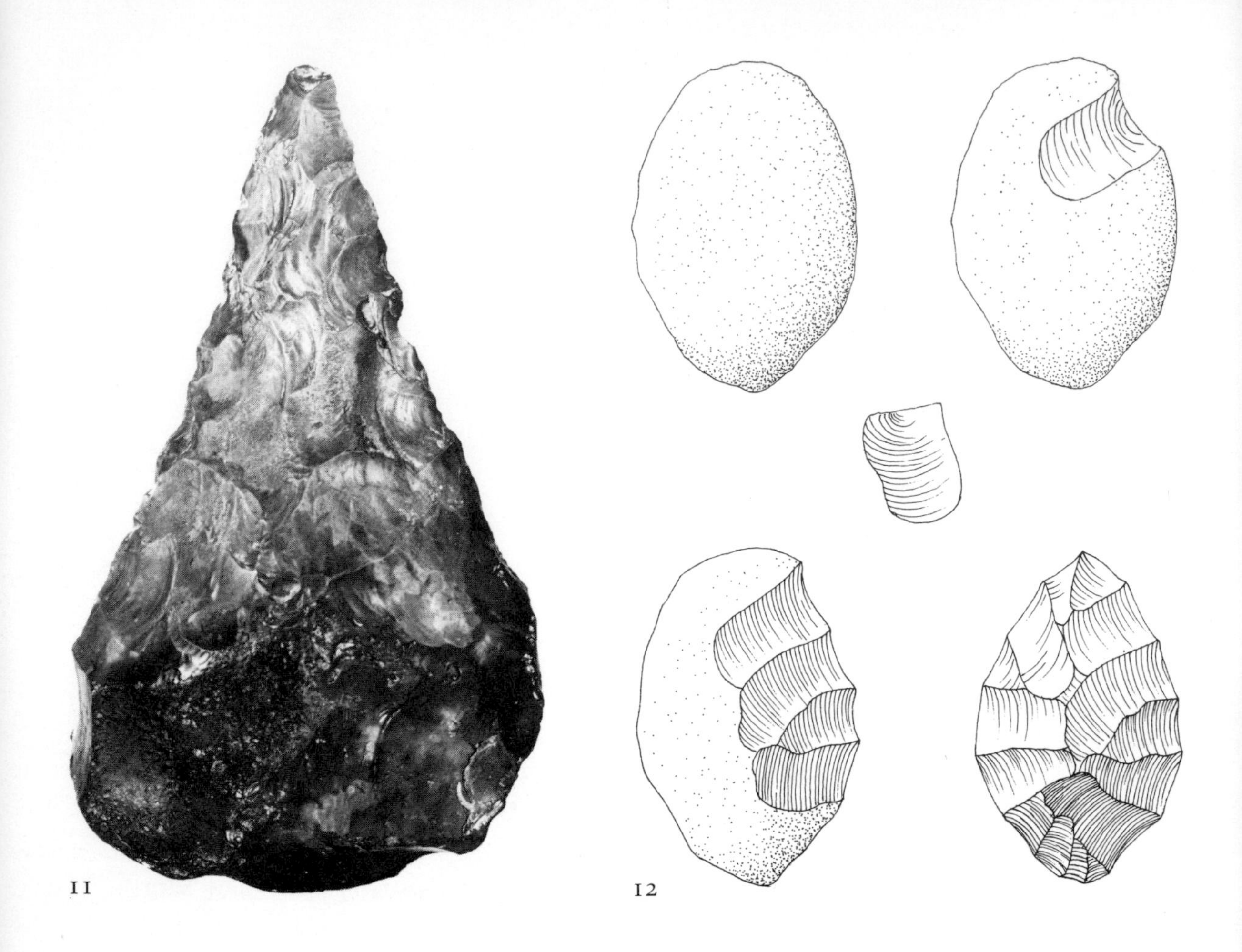
11 12

11 Hand-axe from Gray's Inn, London, about 250,000 B.C.

The first type of tool made by man that we can recognize as such is the so-called hand-axe of flint. It was probably made for no one specific purpose: it may have been used for cutting, chopping, scraping or even digging up roots. This particular implement is of great historic interest, for it was that used by John Frere in 1797 to illustrate the antiquity of man. Hitherto it had been held, as a result of taking the Bible too literally, that the world had been created in the year 4004 B.C. With John Frere and his contemporaries began the study of archaeology as we know it today, and with it the history of early technology.

12 Diagram showing how a hand-axe was made

Hand-axes were made either from flint or from a fine-grained rock that would provide a sharp edge. Beginning with a suitable piece of stone, flakes were hammered off in series around the edge, so that eventually both sides of the tool were covered with scars from which the flakes had been detached. At first the flakes seem to have been discarded, but in time they were used unaltered as small tools while, later still, they were often trimmed to provide knives and scrapers.

was in use it showed little change in form, although gradually the edges were trimmed to an even straighter line and its makers clearly aimed at producing a more geometric form. At the same time the flakes detached during the manufacture of hand-axes were demonstrably used for a number of different purposes, although they were not always trimmed to any specific form.

Whatever its function, the hand-axe must have proved a fairly satisfactory tool, for it remained in use over an enormous period of time. Furthermore, its distribution is curiously wide, for it is found throughout Africa, a large part of Western Asia and in Europe. Looking only at the hand-axes it is tempting to credit early man with very little intellect and, to take an extreme view, to assume that the making of implements became an acquired characteristic much as is the making of nests for birds. Fortunately, however, we know that early man was beginning to master another quite different technology – the control of fire. In the caves and on the river banks where he settled, a sufficiently large number of hearths have been found to show that early man at least a million years ago was building his own fires and maintaining them, although it is impossible to say how he lit them or to what purposes they were put.

At the same time that early man was learning to create neater forms of hand-axe he was also beginning to pay more attention to the flakes struck off during their making, so that by some quarter of a million years ago we find groups of people who were concentrating more on the flakes struck off than on the parent block from which they were detached. The parent block was carefully trimmed in such a way that when a flake was struck off it would be of the required form and could be used directly or with the minimum amount of later shaping. These so-called flake tools were, of course, smaller than the hand-axes and were thus capable of being used for finer work and it is even possible that some of them were set in a haft or handle to serve as spear-heads or knife-blades. It would probably be wrong to dignify these implements with the term 'specialized tools', but at least a step in that direction had been taken.

It will be noticed that in these early periods, so far as stone tools are concerned, only a single stage of production can be detected – starting with the parent block of stone, mankind shaped directly the tool he required. But with the onset of the last glaciation a more sophisticated way of making stone tools was developed. Having learned how to control the shape of the flake struck from the parent block, a system of tool-making was evolved in which long narrow blades of flint with

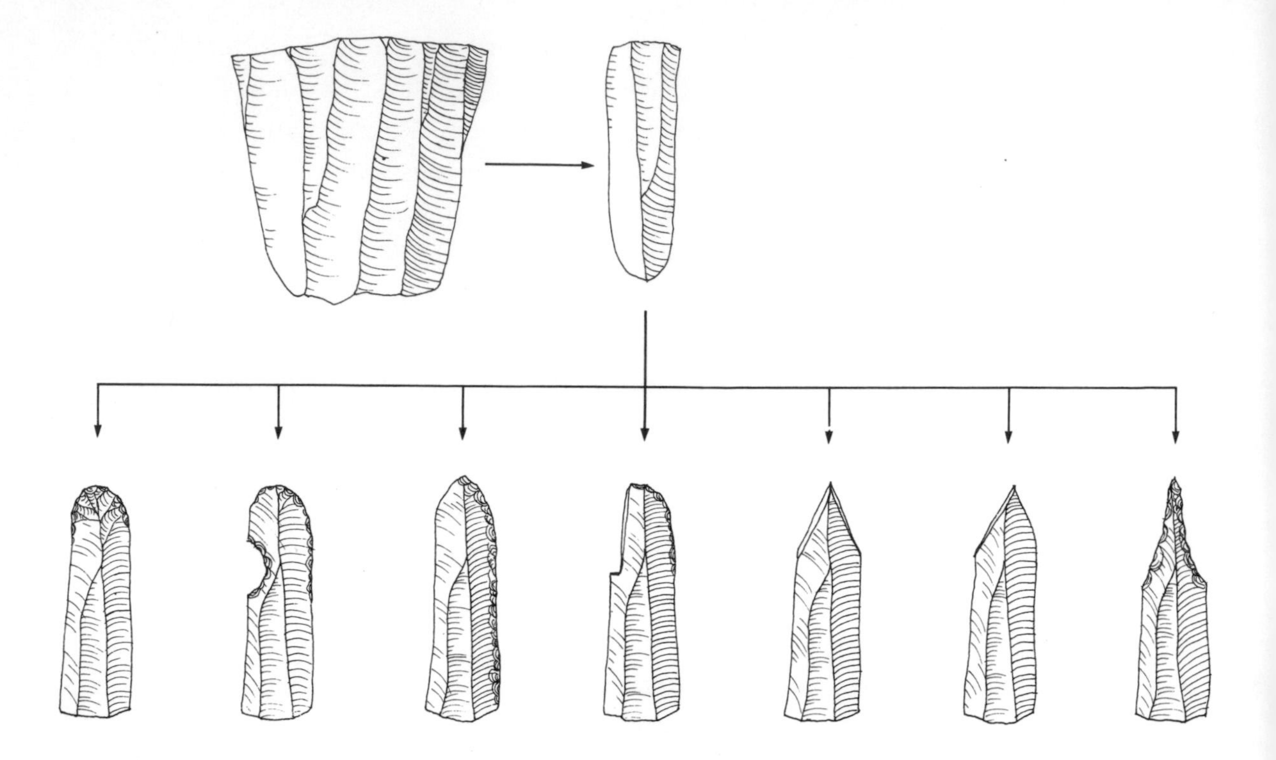

13 Diagram showing how stone blades were made, from about 25,000 B.C. onwards

14 Diagram showing how a flint graver may have been used to carve bone

A new stage in the manufacture of stone tools was reached with the production of blanks from which a number of different tools could be made. The block of stone was so trimmed that from it a series of flat, rectangular pieces could be detached. By further trimming the blanks could be converted into knives, scrapers, chisels or gravers. These tools could then be used to shape other implements of wood, horn, antler and bone.

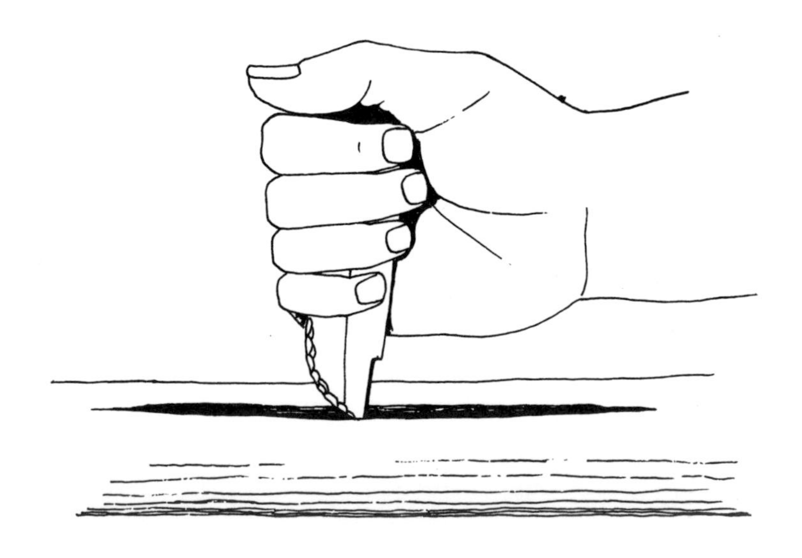

roughly parallel sides were struck in series from the parent core, and from this basic form a wide variety of tools could be made to meet separate purposes. One edge of a blade might be blunted so that it could serve as a small knife; the end of the blade might be rounded to form a scraper; a concavity might be chipped in the edge so that the tool served as a draw-knife; or the blade might be chipped in a variety of ways to provide small chisels or gravers. Thus a new and very important stage was introduced into the making of stone tools. The parent block was no longer worked immediately into the tool, and a flake from the parent block was no longer made directly into an implement. Instead we find an intervening stage in which a blank was produced from which at a later stage tools could be fashioned as required.

Probably our understanding of tools made of materials other than stone used by men during the last glaciation is the greater because of circumstances that allowed better preservation than before, for our knowledge of mankind of this period comes largely from the caves in the limestone rocks of South-west France. In these caves bone and antler objects have been preserved in quantity, as well of course as the many paintings carried out on the walls of the caves. The bone and antler tools show an ever-increasing diversity. Barbed spear-heads and hooks were made for the catching of fish, and bone spear-throwers – long rods with a hook at one end – were used presumably for hunting larger game. From the wall paintings themselves there is sufficient evidence to suggest that the bow and arrow had been developed, although of course the wooden bows have long since perished, while the discovery of the occasional perforated bone needle shows that sewing or net-making was being practised. Simple as they are, the spear-thrower and the bow illustrate that mankind was at least experimenting with elementary mechanical devices.

Fires are a regular feature of caves occupied during the last Ice Age, but whether or not cooking was being practised we do not know. We do, however, find lamps in the form of small bowls made of soft stone in which the fuel was probably animal fat. Indeed, some of the wall paintings were carried out in parts of the caves where daylight could not penetrate, and it has been argued that lamps had to be used in this sort of situation. The pigments used for these paintings, largely charcoal, red and yellow ochre, were pulverized in crude pestles and mortars which, like the lamps, were made of stone. There remains, however, a large field of activity during this period about which we would like to know very much more. We have, for example, no

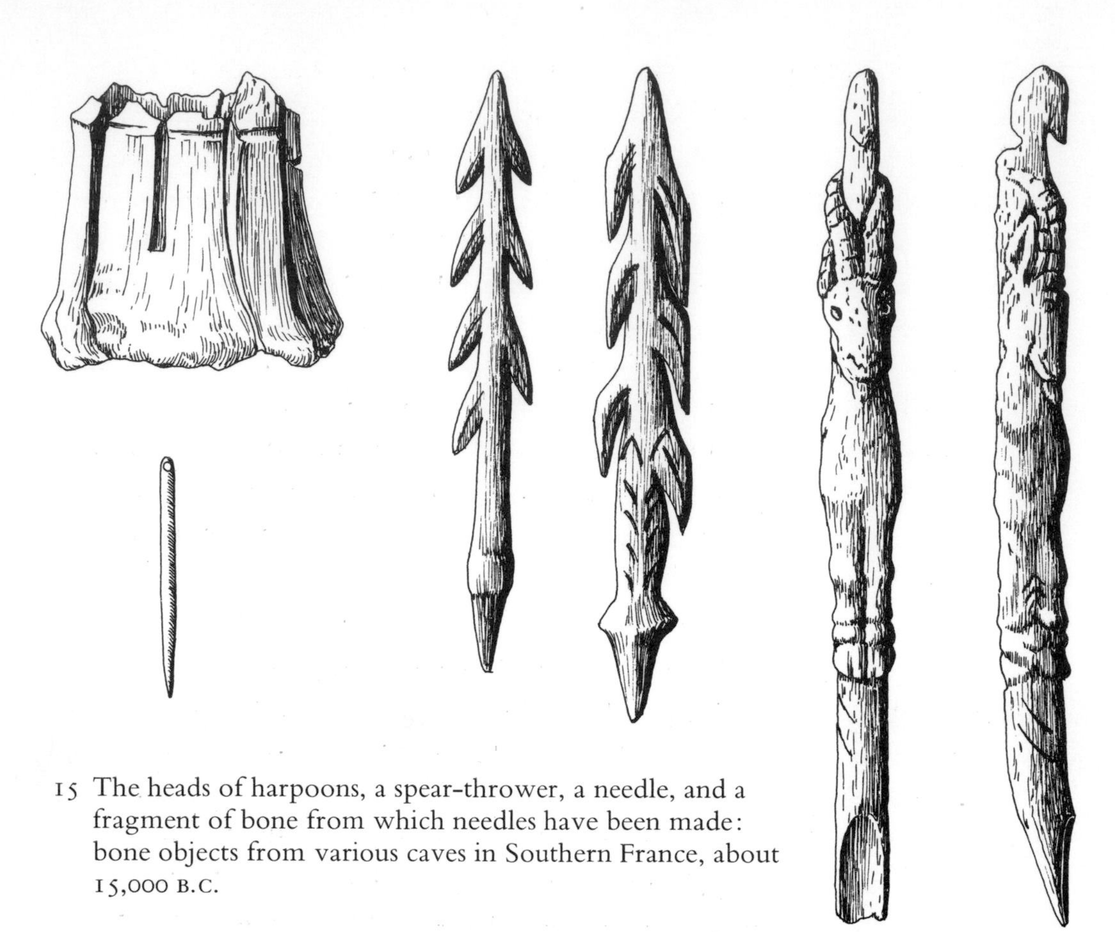

15 The heads of harpoons, a spear-thrower, a needle, and a fragment of bone from which needles have been made: bone objects from various caves in Southern France, about 15,000 B.C.

16 A modern Eskimo using a spear-thrower

Amongst the bone tools used at the end of the Ice Age were barbed harpoons and spear-throwers. The harpoon was presumably set in a long wooden haft. The spear-thrower was manipulated in much the same way as it still is by some peoples today, as for example by many Eskimos. In effect it gave greater length to the throwing arm and thus greater range and power to the missile. The bone needle may have been used either to make nets or to sew skins together.

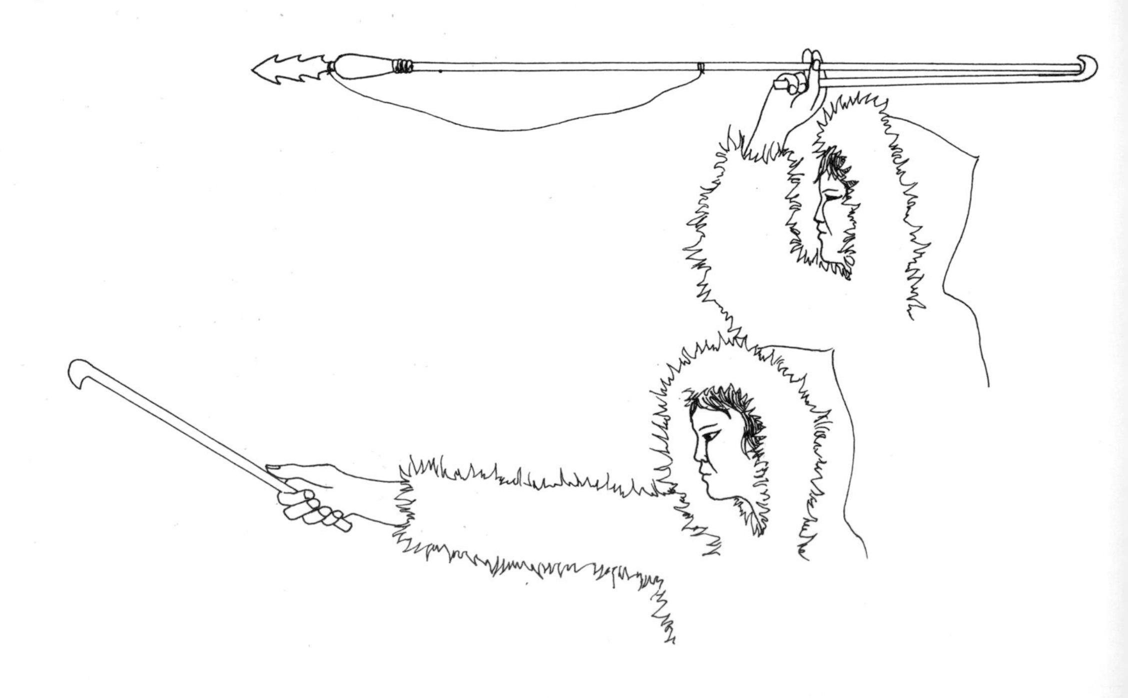

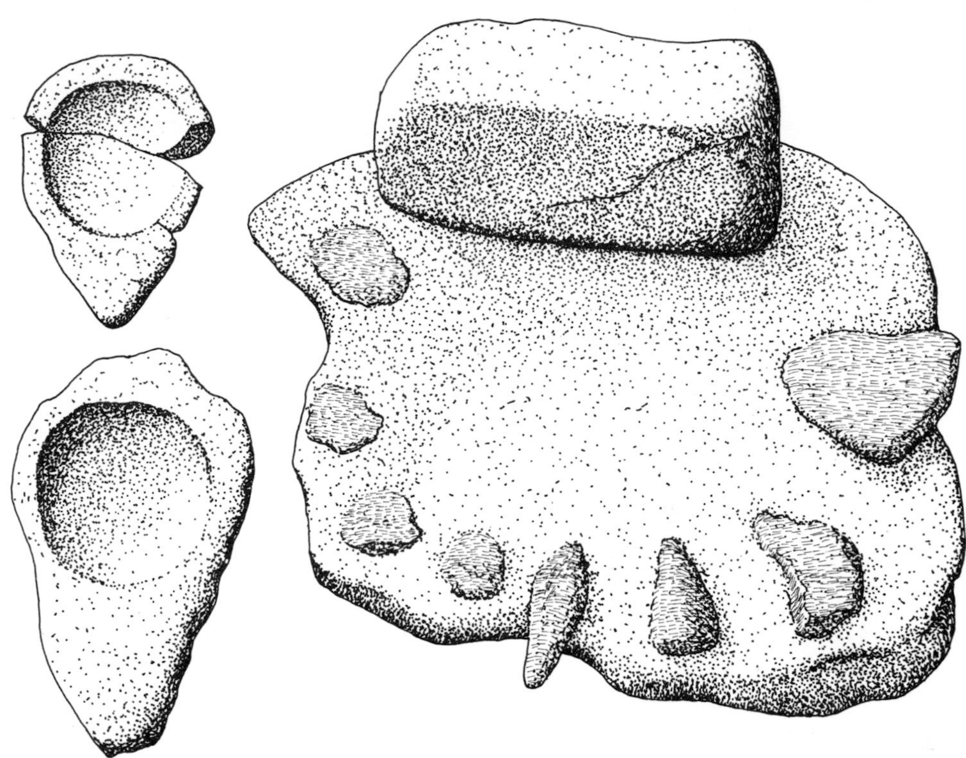

17 Stone lamps, grinding-stones, and pieces of various minerals used as pigments, from caves in Southern France, about 15,000 B.C.

Although man was clearly using fire a quarter of a million years ago, we know of no lamps until the end of the Ice Age. They were made by hollowing a lump of soft stone. Animal fat was probably used as a fuel and a twist of dry plant fibre as a wick. Roman lamps (Figure 235) may seem more sophisticated in shape, but their function was identical.

Pigments used for decorating the walls of caves – charcoal, red and yellow ochre – were reduced to a fine powder in a simple stone mortar using a second stone as pestle. We have no way of knowing whether or not wild grain was being crushed in the same way to make flour.

direct evidence that baskets were being woven, although the presence of the bone sewing needles would suggest that the kind of in-and-out movement required in sewing had probably also been adapted to produce baskets and other woven objects. Furthermore, many of the bone objects are perforated and one is left wondering as to whether this work was carried out by wrist movement as, for example, when one uses a gimlet today, or whether some type of rotary drill had already been developed.

The bone and antler tools were shaped by two principal processes – abrasion and cutting. Blocks of sandstone and other suitable rock were used for polishing the tools, while the cutting was done by using flint knives and gravers, and we can infer, therefore, the devising of tools

18 Bison painted on the ceiling of the cave at Lascaux, Southern France, about 12,000 B.C.

Scholars are unable to agree as to why the caves in Southern France were decorated with pictures of wild animals. In this case it has been suggested that the arrow-like marks on the flanks of the bison had to do with some form of hunting magic. However this may be, two things are certain: the animals were being hunted to provide food, and the painstaking way in which the animals were depicted shows that man was not only taking interest in the shapes of things around him, but was also concerned to reproduce them accurately.

for the making of other tools. Although this development was already implicit in earlier periods, as for example where a stone-scraper had been made possibly for the manufacture of some wooden tool, we can now be perfectly certain that man was planning not one or two but often three or four stages ahead. Thus, to make a bone spear-head, man of the last glaciation would require first to acquire a suitable block of stone; shape it; from it strike off a number of blades; and from these blades manufacture the tools required to shape the bone: while the spear-head itself would have had to be set in a wooden haft which, too, would have had to be shaped, and for which further stone tools would have had to be devised.

It is far from clear, however, whether any of the naturally occurring materials were being modified in other than a mechanical way. We have already noted our ignorance about cooking, but on the whole one is left with the feeling that so far mankind had not begun to experiment with chemical change. Certainly he was using no material that had been so modified that its physical characteristics were entirely altered.

The economy of man during this period was one of a hunter and a gatherer of naturally occurring wild foods, and such equipment as has survived was devised almost entirely for the taking of game. Before we leave the Ice Age, however, we must allow ourselves one further speculation. From the period that immediately follows the Ice Age we know that the dog was already established as a distinct domesticated species, and we can, therefore, legitimately surmise that during the Ice Age dog and man had entered into some form of symbiosis, centred on the hunting activities of the two animals – a relationship in which the dog relied upon man's superior intellect, and the man relied on the dog's superior speed, and in which the dog was rewarded with those parts of the kill unwanted by man.

The ultimate contraction of the northern ice-sheet with the accompanying climatic and floral changes meant that man in Western Europe could no longer hunt his familiar game. Some species, as for example the mammoth, became extinct. Indeed, it is even possible that by the end of the Ice Age the herds of mammoth had been so drastically reduced by man's hunting activities that the species was extinct anyway, while other creatures such as the reindeer, which can survive only where its specific diet in subarctic pine forests is available, slowly migrated northward. Certainly some groups of men moved northward, too, and continued to hunt the old game, but there were others who adapted their hunting methods to the new fauna of the deciduous forest, so that we find Western Europe now peopled by small hunting groups whose game were largely red deer, wild cattle and wild pig, but who also appear to have concentrated more and more on smaller animals and birds as items in their diet. Nevertheless they remained essentially hunters and their technology continued to be essentially that of their ancestors of the Ice Age.

We know very little about life as it was lived in the Near East during the period of the last glaciation. Few sites of this age have been examined and the conditions for preservation are not nearly as favourable as those in the caves of South-western France, but to judge by the few stone tools that we do know, there is little reason to think that either man's way of living or his technology varied greatly from that in Western Europe. On the other hand, during the period of the last Ice Age the countries of the Near East had not undergone a period of extreme cold as had the countries of Central and Western Europe. In all probability the area had been one of high rainfall and the forest cover had been that of deciduous trees, with the result that the game available during this period was largely wild cattle and wild pig, and in the more mountainous areas the wild ancestors of our modern sheep and goats. We shall probably never know how and why the people living in this region took to the domestication of animals and to the deliberate growing of foodstuffs. For both developments numerous theories have been put forward. It seems highly probable that, long before any intentional agriculture, hunting communities had deliberately protected natural stands of wild plants that they found useful in their diet, and may even have fenced them off against wild animals. It is precisely those wild species – cattle and sheep – that would have been attracted by patches of grain-bearing plants that first became domesticated by man. But whatever the mechanics of the change, whatever the cause and effect, somewhere before 6000 B.C. we find a number of widely scattered communities throughout the Near East who had made the change from a pure hunting economy to that of part-time agriculturalist and stock-rearer. It is possible that mankind attempted to domesticate nearly every wild species in his ken and that he succeeded with only a few, the remainder failing to breed in captivity.

The change in basic economy was accompanied by a series of technological advances, some of which may be seen directly as a result of the changed economy and others which seem more likely to have happened as a matter of chance. Thus, in order to grow the new crops – principally wheat and barley – mankind had to clear a space in the forest. This could have been achieved by fire alone, but it would have meant that the larger timber would have been left standing, and it would seem that the axe with the polished stone blade was devised in response to the need to fell trees. Equally, while it is possible to plant grain with no more elaborate tool than a pointed stick, as indeed some primitive people still do today, a forked stick was adapted to serve as a primitive form of hoe.

During the period of the last glaciation the various hunting communities often seem to have created no form of habitation other than a rough shelter, and it could be argued that, because their game was always on the move, they, too, had to live a somewhat nomadic life, and that this state of affairs did not allow them to construct any other

19 Early nineteenth-century wood engravings by Thomas Bewick of (a) 'wild' cattle from English herds kept in parks; (b) the mouflon, one of the varieties of wild sheep from which domesticated sheep were bred; (c) the wild pig; (d) the dingo, the feral dog of Australia

Amongst the animals first domesticated by man were cattle, goats, pigs and the dog. We know very little about the processes by which these animals were tamed and bred, and since we are only aware that they were domesticated by the presence of their bones found during archaeological excavations, we can seldom be certain of their exact appearance. Probably they differed very little from the wild and half-wild species illustrated by Thomas Bewick in 1807.

(a) (b) (c) (d)

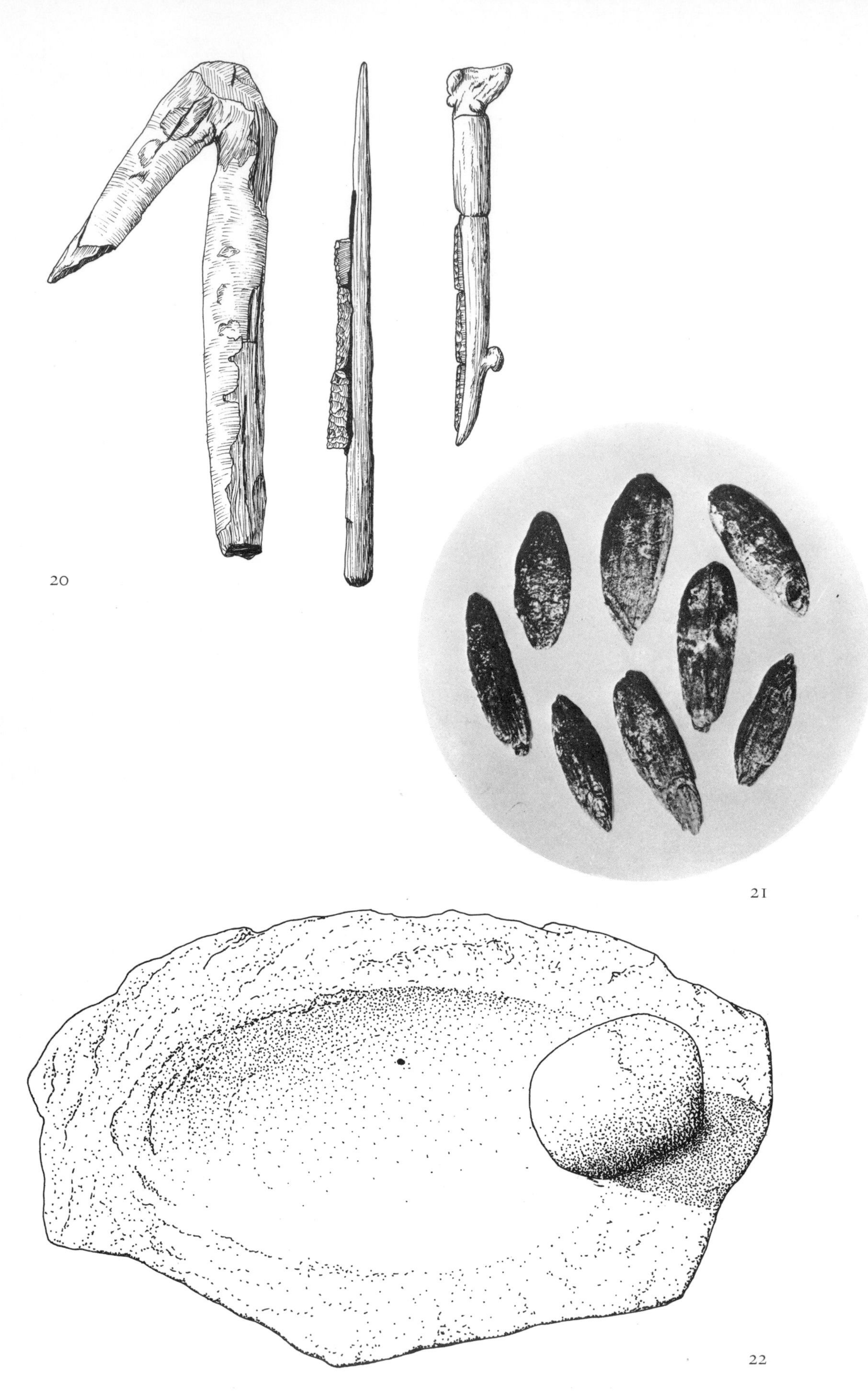
20
21
22

20 A wooden hoe and a flint-bladed sickle with wooden handle from Egypt; and a bone-handled sickle from Palestine, all before 4000 B.C.

21 Surviving grain from Northern Iraq, before 5000 B.C.

22 A hand-mill from Egypt, before 4000 B.C.

23 The impression of a seal from Mesopotamia and a model from Egypt showing hand-mills in use, both before 2000 B.C.

Preserved grain, mostly wheat and barely, as well as the impressions of grains left in pottery, allow us to say that these crops were being cultivated. Less can be said about early agricultural tools. Hoes made of a forked branch may well have been used at this time: they were certainly being used later in Egypt as shown by surviving examples and by wall paintings depicting their use. Reaping was done with sickles made by setting a line of short flint blades into a bone or wooden handle. The grain was reduced to flour using a flat lower millstone and a rounded rubber. With use the lower stone became hollowed.

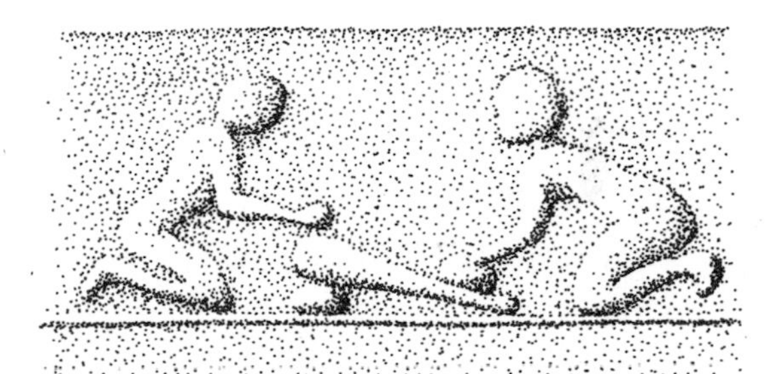

23

type of building. On the other hand, the deposits, often very deep, in many of the caves would suggest a long occupation, perhaps only seasonal, but certainly not nomadic. Moreover, there are a few known instances, both in Central Asia and the Near East, where substantial villages had been built long before there were any manifestations of intentional farming. Of these Jericho is undoubtedly one of the most widely known. With its excellent spring of fresh water and its dominating position over an easy crossing of the River Jordan we need not be surprised that early hunting communities chose this spot as a rendezvous. Furthermore, many primitive communities that still live today largely by hunting are often more sedentary than one might suppose, for the products of the chase, such as hides and meat, frequently require long periods for their processing, especially where simple methods of curing, either by salting or smoking, are involved. It is, of course, pure speculation that the process of smoking was one of the reasons for early man's need for fires, but the possibility should not be ignored. The fact remains that small, crudely built villages clearly existed long before man put his hand to farming.

With the development of agriculture and stock-breeding the technique of house-building seems, nevertheless, to have developed very rapidly. In view of what has already been said about the need to clear the forest we might have expected the earliest houses to have been made largely of timber, but in fact this is not so. Throughout the Near East often the earliest domestic building material was sun-dried brick, made from alluvial mud, hand-modelled and oval in section, with a strong resemblance to the loaf of bread known as a bloomar. In early villages on the Iranian Plateau and in Northern Mesopotamia and in the lower levels at Jericho this type of brick was used for building, and one is tempted to suggest that their makers were more familiar with structures made of rounded stones. To a certain extent this view is borne out by other early settlements in areas in which stone of suitable sizes and shapes was used for the construction of early houses. On the other hand, when we look at the earliest known settlements on the Anatolian Plateau we find a very different picture, for here, although sun-dried brick was the common building material it was evidently not hand-modelled. The bricks are too uniform and too regular to have been produced in anything other than a mould. What the moulds looked like we cannot say, although many hundreds of years later we have adequate Egyptian wall paintings of bricks being made in rectangular moulds: nor can we say how or where the idea developed. It is very tempting to suppose that the people using these moulded bricks were

24 The stone foundations of huts from a village in Northern Iran, before 5000 B.C.

In mountainous areas, where rock was plentiful, houses were often built of stone and rammed earth or pisé. The photograph shows the excavation of a village at Jarmo in Northern Iran. The stone footings of the houses can be seen clearly, as well as the clay-built oven (top centre). It is impossible to say with certainty how high the walls were, or how the houses were roofed, when in use.

initially familiar with a rock that broke naturally into a flat tabular form, but this would be pure speculation. It would be more useful to notice that before 6000 B.C. in Anatolia at least a simple moulding device had been invented.

The people who made their houses of mud-brick did not use the material solely for this purpose, for the walls of their buildings were normally plastered with a layer of mud while the floor was made of a carefully laid layer of clay. In its plastic state clay was also used for the modelling of such things as small figures and toy animals. It was also discovered during this period that, when dry, the material could be hardened by firing, and the production of pottery is often taken to be one of the more remarkable inventions made by man during this epoch.

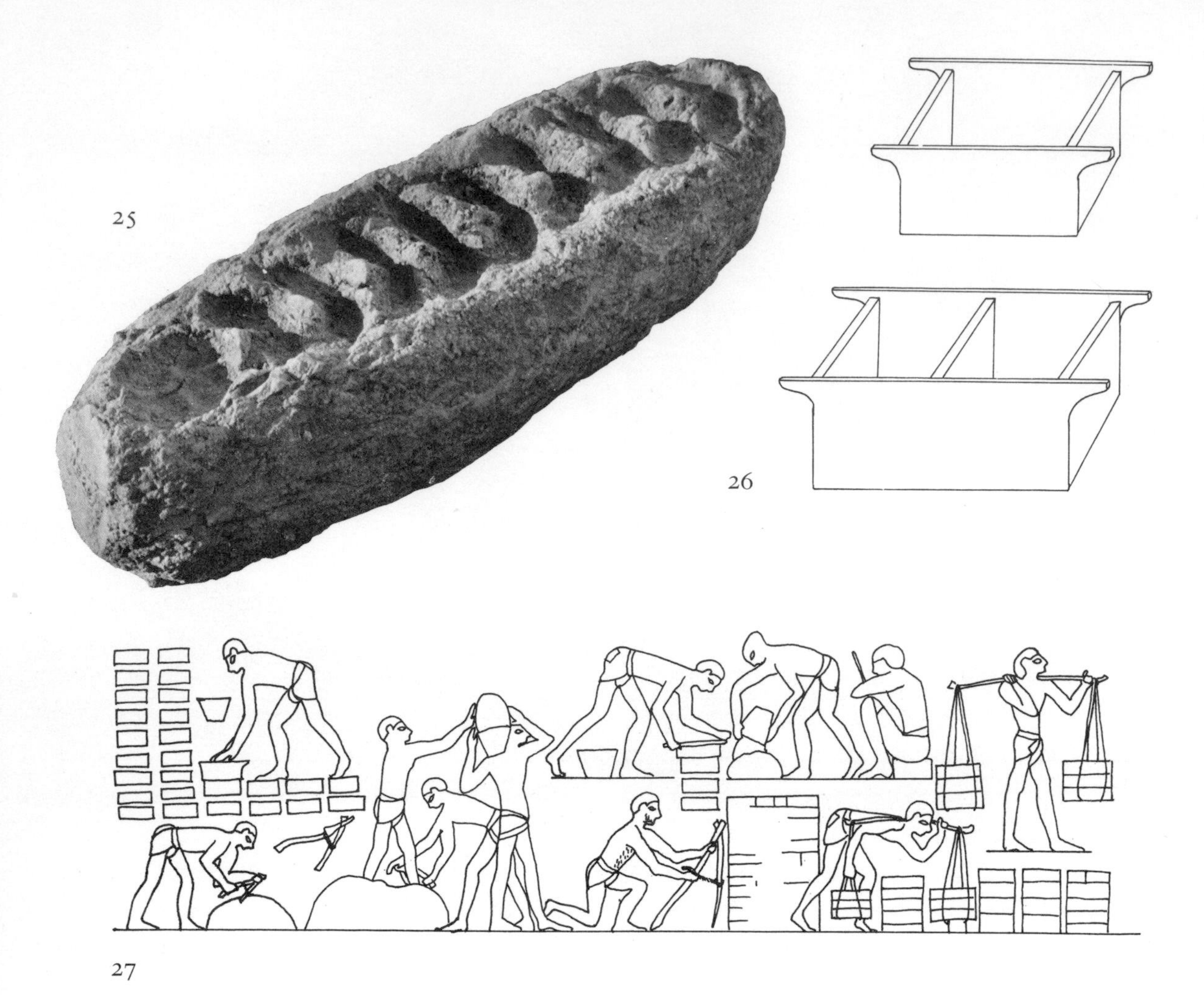

25 Hand-made mud-bricks, Jericho, before 5000 B.C.

Even in areas where stone was plentiful it was not always used, and throughout the Near East sun-dried mud-brick was very commonly used for building. In the earlier stages the mud was modelled by hand, in this case to give bricks shaped something like a bloomar loaf. More mud was used as a mortar to join brick to brick when building.

26 Diagram showing the form of single and double brick-moulds used in antiquity

27 Egyptian brick moulds in use, from a tomb painting of about 1500 B.C.

Before 5000 B.C. sun-dried brick was being made in moulds in Crete and Western Turkey, and although one does not know the exact details, the moulds were probably bottomless wooden boxes with handles for lifting. An Egyptian wall painting of a much later date shows how the process was carried out. Mud was puddled with water and packed down into the moulds, which were then lifted away, leaving the newly formed brick sitting in the full sun to dry. The process is still used in exactly the same manner throughout the large part of the Near East today.

28 The impression of a seal showing what is probably a reed or wattle hut, Mesopotamia, about 3000 B.C.

29 A modern rush-matting hut from Northern Nigeria

In the river valleys, particularly in the deltas, where reeds were readily obtained, they were probably used for building purposes, although they have long since rotted away. On seals of a much later date huts are often depicted which were clearly made from bundles of reeds lashed together, or from reed matting. Similar buildings are still being made in Northern Nigeria and by the Arabs in the Euphrates delta.

28

29

30 Pottery from Egypt, before 4000 B.C.

Containers made of various materials must have been used by man even during the last Ice Age, but these were probably all made of organic materials such as hides or basketry. From a very small number of excavations where conditions of preservation have been particularly favourable it is clear that early farming communities were making vessels of wood, and it is probable that this material, too, was being used at an even earlier period. In other areas where suitably soft rocks were to be had we find that early farming people made bowls of stone; while gourds and similar fruits provided mankind with a number of containers of limited shapes and dimensions. Looking only at the rather simple forms of early pottery archaeologists have often concluded that the wares imitate vessels made of these different materials, and that pottery to early man was essentially an *ersatz* material. Nevertheless, even the earliest pottery that we know shows certain sophistications. In the first place, the clay was always mixed with some other material such as sand or crushed rock or even materials of organic origin, partly to prevent excessive shrinkage during drying, and partly to make the clay less prone to break during firing. Furthermore, it is seldom that we find a single type of clay mixture used for making all the pottery at any particular place, and it would seem that a great deal of attention was being paid to the properties of the finished product. Thus, we may find on the same site porous pottery that would have served well for keeping water cool, and less porous pottery that would have been more suitable for cooking vessels, while the surface of many pots was brought to a high shine by burnishing with a stone to render it less porous still. The same technique of burnishing was on occasions, equally, used to polish the floors and even the walls in early houses. Indeed, the technique seems to have been used for this latter purpose first and only applied later to the smoothing of pottery.

The usual method of shaping vessels seems to have been, first, to mould the base of the pot over some suitable hemispherical form, often one supposes the base of an existing pot, and then to add rings of clay to this so as to extend the height, a method still widely practised in backward parts of the world today. When dry, the pottery was probably fired either in the domestic hearth or in a bonfire especially made for the purpose, for we have no evidence of the use of kilns at this early date.

As with so many early inventions, it seems unlikely that we shall ever be able to retrace the steps by which pottery came into being and it is probable that at the back of the invention lay a long period of experimentation with vessels made of unfired clay. Nevertheless, the appearance of pottery at a time when grain-bearing plants were first being domesticated and needed to be stored, and when the grain itself had to be cooked, must be far from coincidental, while the different formulations used for the material from which the pottery was made suggest that, to its users, it was anything but an *ersatz* product.

The grain crops gave rise to a number of different specialized tools. One of these was the sickle, usually made by setting short lengths of flint into a suitable organic handle, and thus we find for example sickles made by first removing the teeth from the lower jawbone of an animal, and then inserting short lengths of flint blade into the sockets, using resin as an adhesive. To make the foodstuff more palatable it was necessary to reduce it to a flour, and we frequently find primitive mills made from two stones – a lower flat piece of sandstone upon which the grain was placed and an upper elongated pebble of the same rock under which it was crushed. With use, of course, the lower stone became hollowed and the lower surface of the rubbing pebble became flattened.

31

32

33

31 Men making pots on the shore of Lake Victoria

32 Men making pots in Northern Nigeria

The discovery that clay vessels could be fired to produce the more durable pottery also belongs to this period. Methods of manufacture were simple, requiring no specialized pieces of equipment. The clay was mixed with sand or crushed rock and shaped by hand over an existing pot or built up ring by ring. Many peoples still make their pottery in this manner today. Shapes were simple and often seem to imitate skin containers or gourds, both of which may well have been used for storage and drinking vessels long before pottery was created.

33 Firing pottery without a kiln, Western Turkey

Pottery was fired either in the domestic hearth or in special bonfires. Many people still fire their undecorated wares in this way today, and the method need not have been as unproductive as one might think. In the village in Western Turkey seen in this photograph, although there are only fifty families, the women manufacture many thousands of cooking pots a year.

These simple mills were, in effect, only an adaptation of the mortars used by man during the Ice Age to prepare his pigments. One is left wondering to what other purposes the earlier hunting communities may have put their mortars, and whether wild seeds were even then being prepared for cooking.

Sandstone rubbing-stones were equally used for the polishing of axe-heads and other tools made of hard, fine-grained rock. At first sight it seems rather unnecessary to have bothered to bring a stone axe-head to a fine, uniform polish, but experiments carried out in Denmark and elsewhere in which trees have been felled using stone axes have shown that the polished axe was far more efficient than an axe-head made simply by flaking the material roughly into shape.

Mud for the making of bricks, clay for the making of pottery and slabs of sandstone for various polishing techniques are fairly ubiquitous, but the same cannot be said for the rocks required for the making of cutting-tools, and we find during this period very definite evidence of trade in these raw materials. For a people living deep in the interior of an upland plateau or an alluvial plain the fine metamorphosed rock required for making heavy tools such as axes was often only to be found in some distant mountain range, and we can only guess at the process by which it came to be brought to the village. Nevertheless, outcrops of this type of rock are fairly commonly distributed throughout the whole of the Near East. For the making of cutting-tools by

34 Part of an Egyptian tomb painting showing stone blades being flaked, about 2000 B.C.

35 A fine example of a pressure-flaked blade from Egypt, about 3500 B.C.

Weapons, knives and other cutting tools were made of flint, obsidian, or any rock that would fracture to give a sharp edge. The tools were carefully shaped, and thinned by the removal of a mass of small flakes from the surface. This was presumably done by pressing down on the object being shaped with a point made of bone or hard wood. A later Egyptian wall painting shows men trimming tools in this way.

36 Polished stone axe-heads from Egypt, before 4000 B.C.

Many stone tools, especially axes, were given an overall polish by rubbing up and down against a piece of sandstone similar to that used for making corn mills.

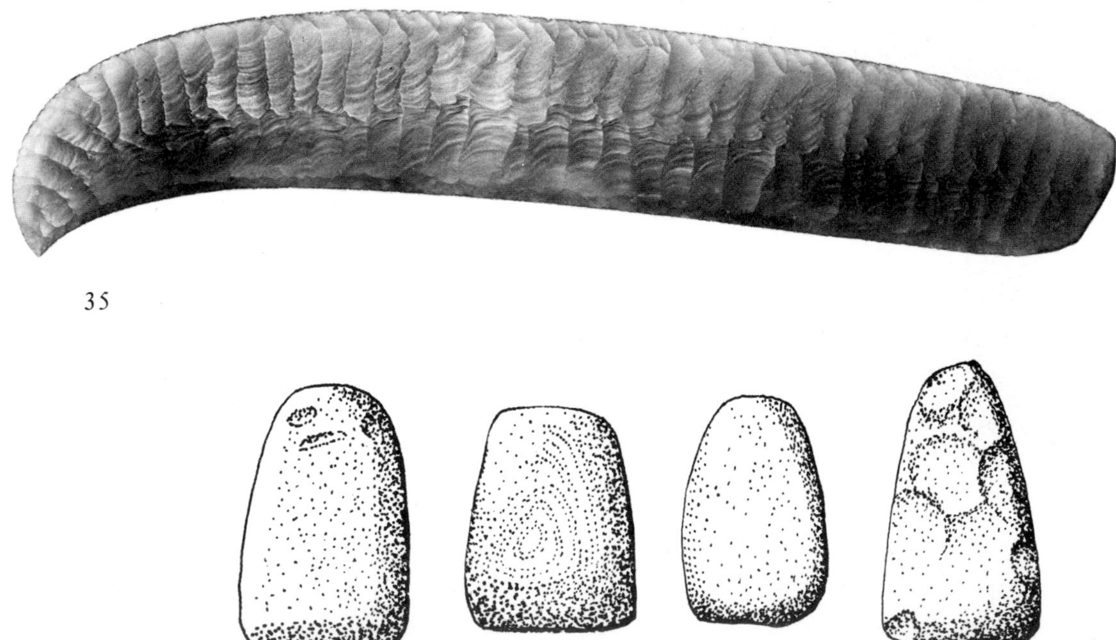

35

36

the older technique of flaking, on the other hand, the naturally occurring glass of volcanic origin, obsidian, was not only greatly favoured but also in short supply, for the number of suitable deposits of this material were very limited, and recent research has shown, for example, that obsidian was being traded from a number of deposits in Eastern Turkey to villages in Iraq, Mesopotamia, Palestine, Syria and Western Anatolia. No matter how the trade was conducted, it undoubtedly helped to open up the routes of communication between the various rather isolated village groups.

The method by which flint and obsidian were flaked during this period seems to have undergone a very radical change from that in vogue during most of the Ice Age, and there is no very obvious reason why this should be so. Tools were still made largely from the parallel-sided blade as during the last Ice Age, but the tools were considerably thinned down, and made far more geometric in form by the removal of a vast number of very small flakes from the surface. This is very difficult to achieve by merely striking with a stone, and it seems almost certain that these very fine flakes were removed by pressing against the edge of the flint or obsidian blade with a point of bone or hardened wood, a process often referred to as 'pressure-flaking'. The technique was not newly discovered during this period for it was briefly anticipated during the last Ice Age, but as a method of manufacture it was then short-lived. Among the farming communities, however, it became universally adopted throughout the whole of the Near East. Pressure-flaking is undoubtedly a knack, and a fairly difficult one to master at that. It is questionable whether it was adopted by every member of the community or merely by a chosen few. The use of the technique, however, allowed a far greater precision in the shaping of tools than hitherto, and thus, where previously arrow-heads, for example, were made by inserting a fairly irregularly shaped fragment of a blade into a shaft, we now find arrow-points of greatly varying sizes, some with and some without barbs, and it looks very much as though mankind was now devoting a very great deal more thought than before to the shape and finish of his tools and weapons to make them more efficient.

So far we have spoken almost entirely of the more durable materials, but for the use of wood and other plant materials we must rely upon inference and a great deal of speculation. For example, some stone tools have neatly made cylindrical holes for hafting and these could indeed have been made by rotating a solid point of wood or bone between the palms of the hands, keeping the work lubricated with water while using sand as an abrasive. On the other hand, we know from the

37 A complete and a fragmentary mace-head from Egypt, before 4000 B.C.

38 Diagram showing how the bow-drill was used

39 The Egyptian use of the bow-drill as shown in a hieroglyph from a tomb of about 2000 B.C.

Stone beads and sometimes axe-heads were perforated by means of a bow-drill. The point of the drill may have been either solid or made of a section of hollow bone; sand acted as an abrasive to cut away the rock, and water was used as a lubricant. The process can, again, be seen in a later Egyptian wall painting. The broken mace-head shows clearly that it was drilled first from one side and then from the other.

40 A basket from an Egyptian burial, before 4000 B.C.

Baskets may well have been made long before the end of the Ice Age, although none have survived. By 5000 B.C. the art of basket-making was already highly developed. Fragments of cloth show that both spinning and weaving were being practised, but nothing has been discovered so far to show what kind of loom was in use.

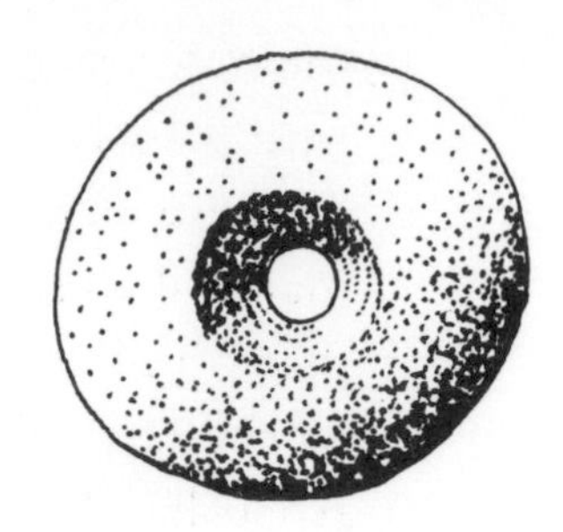

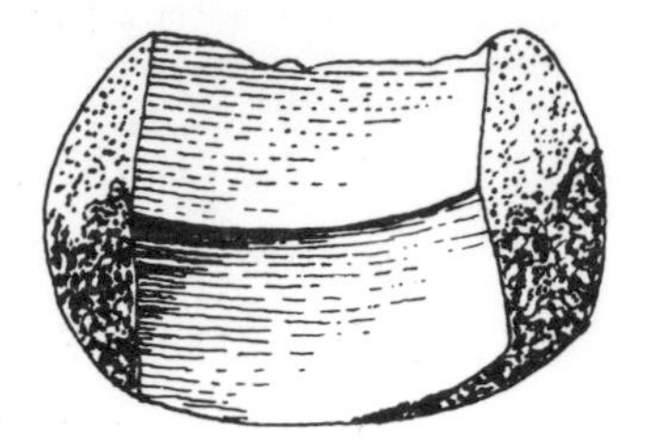

37

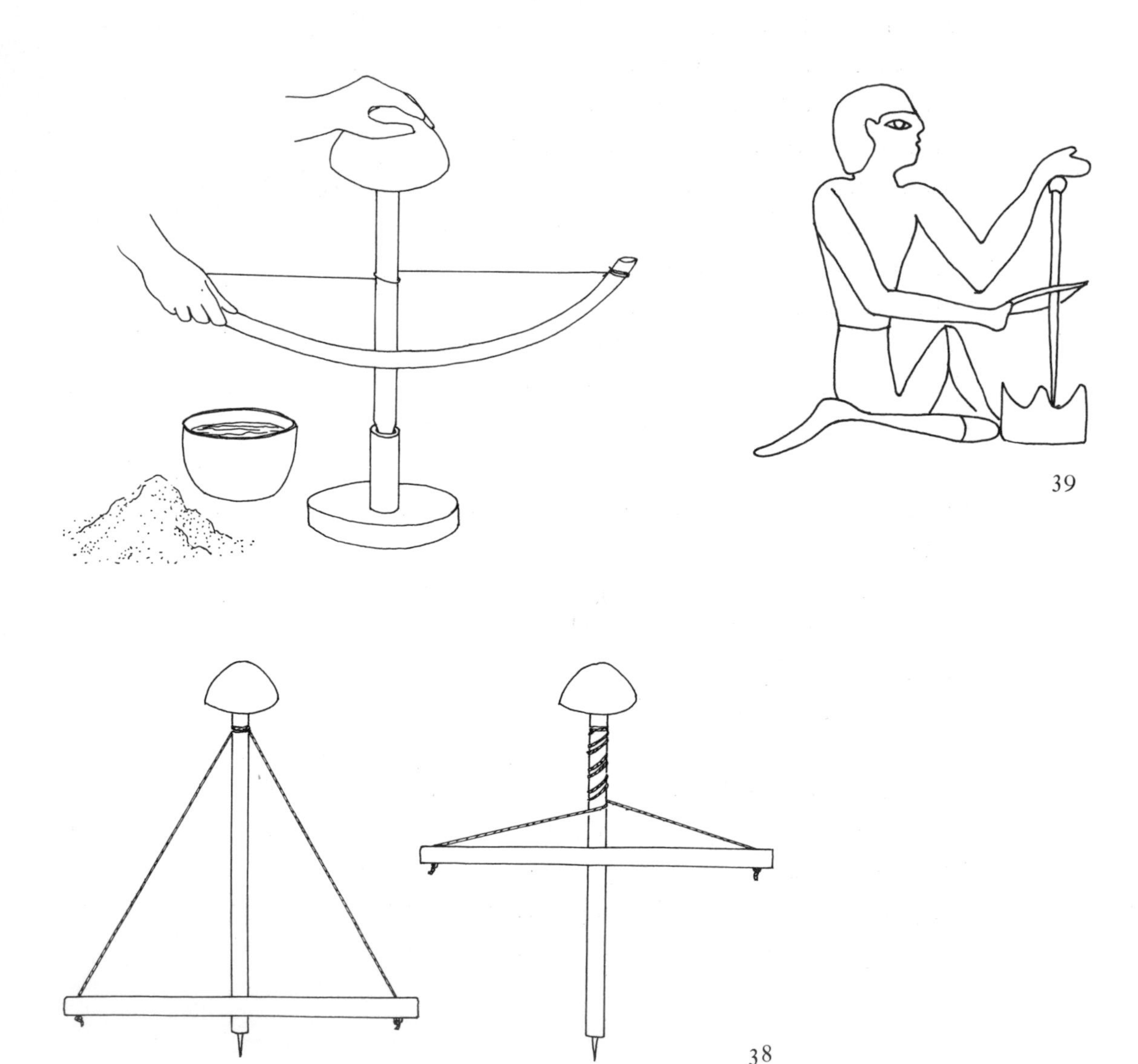

39

38

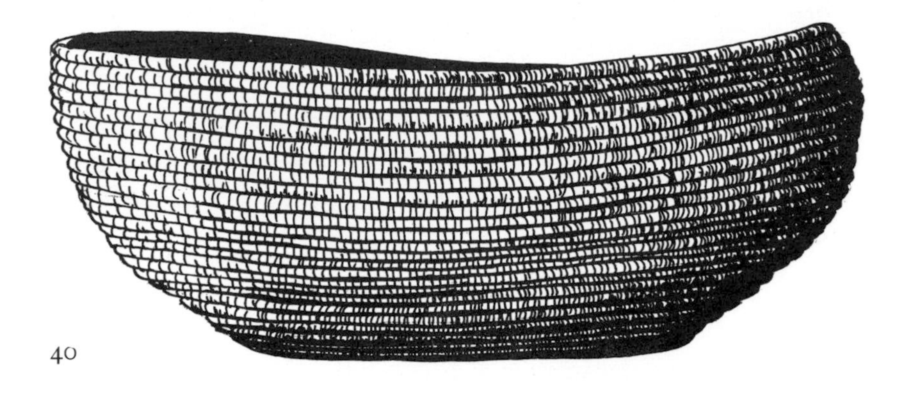

40

arrow-heads and from a very limited number of wall paintings that the bow was very commonly being used by these people, and it is therefore very tempting to suppose that they had already tumbled upon the idea of the bow-drill in which the drilling point, twisted into the string of the bow, was rotated backwards and forwards.

We know that mats and baskets were being woven from such things as reeds during this period, for a very small number have survived, and we occasionally have the impressions of matting left on a clay floor or on the base of a pot. It is not certain, however, when cloth textiles began to be produced for we have no very early impressions or remains. Nevertheless we can be pretty sure that spun thread was being manufactured. From a number of sites have come spindle whorls, the small fly-wheels that fit over the end of the spindle used during the making of thread. Of course thread can be made without any equipment at all by twisting the strands of fibre between the thigh and the palm of one hand while even a simple spindle need have no whorl, for there are primitive people today who have a spindle in which the function of the fly-wheel is performed by a cross of wood. One supposes, therefore, that the clay or stone spindle whorl is a fairly late development in the manufacture of thread. But it is of course one thing to make thread and quite another matter to weave a fabric from it, and thus while we may guess that occasionally textiles were being made at this early period we have no direct evidence. Equally, since no leather or hides have survived, and we can recognize no specific leather-working tools, our ignorance about the use of these materials at this period leaves an awkward gap in the record, although we may surmise that nearly all animal remains were being used for one purpose or another.

There is one further enormous gap in the record about which we must say something. At Knossos, in Crete, in the lowest levels of building which must have taken place before 5000 B.C., there dwelt a people with all the aspects of early farming that we have just outlined – a people who, incidentally, like their neighbours in Anatolia, were moulding their mud-brick. These people, therefore, must have come to the island by some form of ship or raft. What these early sea-going craft looked like we cannot even begin to guess, and indeed it is not until many hundreds of years later, when we find models of boats and wall paintings, that we have any idea of what early boats looked like, or indeed of what they were made. Nevertheless, we are probably justified in supposing that the craft must have already been fairly well developed to allow this sort of sea passage even in fine weather, and that in the rivers and lakes of the mainland, therefore, mankind had already taken to messing about in boats, or on rafts, and that some river transport, even if rather limited, was being used.

By roughly 5000 B.C. early farming communities of the type just described had been established over a somewhat limited area of the Near East – from the more favourable valleys of mainland Greece in the west, through Southern Anatolia, in Eastern Turkey, Syria and south into Palestine, in the Mesopotamian Valley and in the upland valleys and on the plateaux of what are today Northern Iraq and Persia. As we have just seen, on some of the larger Mediterranean

41

42

41 Reed-rafts in use in the lower Euphrates Valley

42 Hollowed log canoe and raft, the River Amazon, Peru

The fact that people were moving from one island to another in the Mediterranean tells us that some kinds of craft were in use. In the river valleys rafts made of reed may well have been employed, but in the Mediterranean it seems more probable that substantial log-rafts were used, or even canoes made by hollowing out large logs.

All three types of vessel are still in use today, as for example in Peru, where the canoes and reed boats are used mainly as personal transport, while the log-raft is more commonly used for shifting heavy burdens.

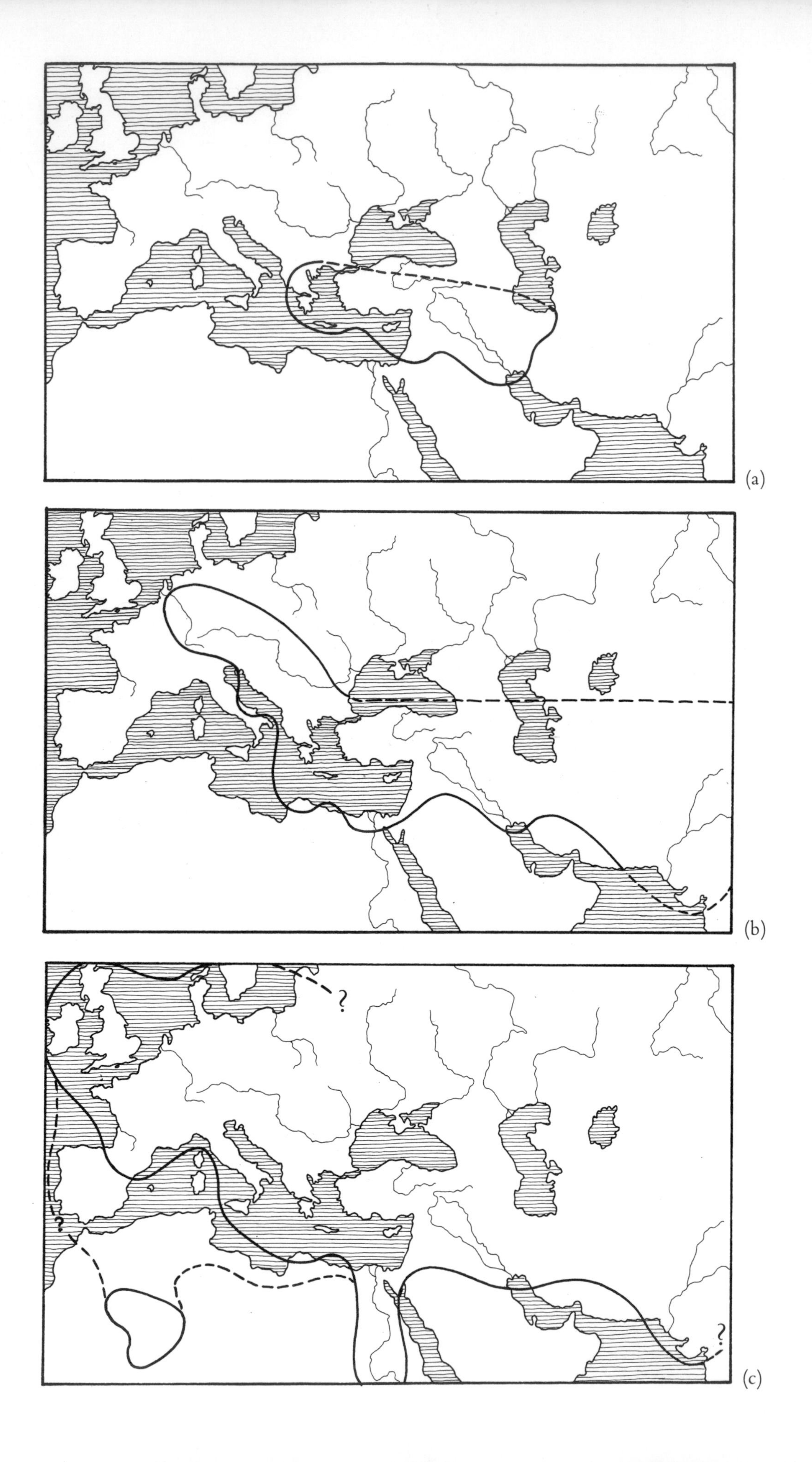

(a)
(b)
?
?
?
(c)

43 The spread of farming: (a) up to 5000 B.C.; (b) 5000–4000 B.C.; (c) 4000–3000 B.C.

By roughly 5000 B.C. villages in which agriculture and stock-rearing were being practised were established throughout the Tigris–Euphrates Valley, the Levant, Anatolia, mainland Greece and on at least one island in the Eastern Mediterranean, Crete. A thousand years later, by 4000 B.C., farming had spread eastwards, perhaps as far as the Indus Valley; south into the Lower Nile and North African coastal belt; and west, up the Danube Valley into central Europe. Within the subsequent thousand years agriculture and animal husbandry were general throughout Europe, and had spread throughout the length of the Egyptian Nile and even into the larger Sahara oases.

These maps were compiled by plotting the earliest known agricultural settlements in each region. Uncertainties exist about some areas, as for example the Iberian Peninsula, due to lack of archaeological investigations that could throw light on the subject.

islands such as Cyprus and Crete similar communities had come into being. It was from this somewhat diffuse nucleus that the practice of farming with its accompanying technologies spread slowly throughout Europe into Africa and Northern India during the next two thousand or so years. For the layman, often accustomed to thinking of Egyptian civilization as having developed particularly early, it is very imporant to notice that by 5000 B.C. the people of the Nile were still living as primitive hunting communities, for later we shall see the same pattern of events repeated. The idea that early Egypt achieved a higher state of technological skill than other countries of the Near East is purely illusory and stems from the better preservation of many of the materials in that country than elsewhere. More commonly than not, we shall find that Egypt received her technologies from her neighbour countries at a rather later date than elsewhere.

Communities founded on the farming of a very limited number of crops and the domestication of a few animals, with technologies based almost entirely upon stone, clay and organic materials, are of course still with us today in many parts of the world difficult of access as, for example, the Basin of the Amazon, particularly the upper reaches of the river, and in the highlands of New Guinea. On the whole these people live happy and contented lives. They have reached, as it were, a level in technological advance and there is little spur to make them change. We shall now have to consider why it was that, shortly after 5000 B.C., in the Near East we suddenly find a host of technological advances that were ultimately to affect the whole future of mankind.

3

Noah's Sons

(5000–3000 BC)

It has sometimes been suggested that the wall paintings carried out by the hunters of South-western France in the caves during the Ice Age are unusual and represent a flowering of the arts which died with the passage of the last glaciation. This is now known to be not completely true, for the early farming communities that we have described in the last chapter frequently carried out quite elaborate paintings on the walls of their houses. The style is certainly different as, too, may have been the reason for carrying out these paintings, but the two groups share in common a feature that we are often apt to overlook – a certain gaiety. Material recovered from excavation is usually broken, weather-worn and grubby and we could be forgiven for imagining that early man lived a somewhat shambling and dirty life surrounded by objects of grey and gloomy browns and reds, but the people of the Amazon and New Guinea who today live the same primitive life should warn us against this idea, for they are apt to decorate all manner of things with vivid designs carried out in the brightest colours that nature will provide. Furthermore, the early farmers were not content only to decorate their handicrafts; they appear to have decorated themselves as well, and we find small pestles and mortars used for grinding the pigments in the making of cosmetics on many of their sites. This, of course, meant that a search had to be made for suitable pigments, and this search ultimately led mankind to the ores of at least two metals. Yellow ochre or limonite, and red ochre or hematite, commonly known as jeweller's rouge, are both ores of iron, while the green mineral, malachite, and the blue mineral, azurite, are both ores of copper. But while iron never occurs as the metal in ore deposits the same is not true of copper, and occasionally sizeable lumps of metallic copper are to be found amongst the ores of the metal. It is, therefore, more than just a possibility that mankind's first interest in metallic copper was aroused while he was in fact searching for a suitable green pigment with which to make eye-shadow.

44 A cosmetic pestle and mortar from Northern Iran, about 4000 B.C.

45 Men preparing for a festival, Kenya. They are using green, blue and red earth pigments

Brightly coloured minerals – red and yellow ochres, and the blue and green ores of copper – were used as cosmetics. The pigments were ground to a fine powder in a small stone pestle and mortar and then probably mixed with animal fats to make rouge and eye-shadow. It was, perhaps, the search for blue and green colours that first led men to deposits of metallic copper found in small quantities amongst the ores.

46 Part of an Egyptian tomb painting showing metal-workers, about 2000 B.C.

Both gold and copper objects were at first shaped only by hammering small pieces of metal that were found to occur naturally. For this purpose rounded pebbles were used, and they continued to be employed as hammers for many centuries thereafter, as can be seen in later Egyptian tomb paintings. At first the metal was heated in the domestic hearth, but later small forges were used in which the fuel was brought to an adequate temperature by the craftsmen blowing into the fire through tubes of wood or reed.

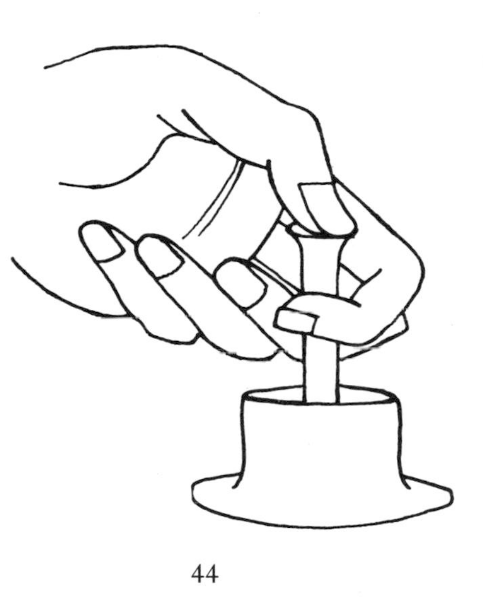

44

45

46

Apart from its colour and lustre metallic copper had a number of properties that mankind had so far not encountered. Although it could be polished like any normal stone, using blocks of abrasive, it could not be chipped as could, for example, flint or obsidian or any of the finer-grained rocks, and attempts to chip it would show that it would bend and stay bent. It could in fact be hammered into the required shape. Thus the earliest metal artifacts that we know are small pieces of copper that have been hammered into various forms, generally rings and other small ornaments. They were small because the earliest pieces of copper available were themselves small and mankind had found no way of joining the fragments. Furthermore, the amount of hammering that could be used in shaping native copper was severely limited, for if hammering were carried beyond a certain point the metal became brittle and ultimately cracked, and indeed modern metallurgical examination of some of these very early metal objects shows that they had been hammered to a point very near to that at which they could be expected to break up. However, the stresses that build up within the metal to cause the cracking can be relieved by heating the metal in a fire to a fairly high temperature, at which it begins to glow red. On cooling, the metal can then be hammered again to shape it until it begins to harden, when it must be heated once again to relieve the stresses that have built up. This process of relieving the stress in metal by reheating is today known as annealing, and it would seem that within a century or two of the first use of copper the idea of annealing had already been discovered. The discovery was, indeed, a very important one, chiefly because it introduced mankind to the concept that metals can be altered by the use of high temperatures. We could argue that the discovery of annealing was purely accidental – that some fragment of the metal got dropped into a fire and afterwards was found to have become softened – but it could be argued equally that it was the result of a logical process. Many natural materials as, for example, bitumen or resins soften with heating, and it seems therefore highly probable that the use of fire to relieve the hardness of hammered copper may well have been an attempt to apply the principle of softening by heating.

The occurrence of native copper can never have been widespread, for not only is the metal rare in ore deposits, but the ores of copper themselves are very limited in occurrence. Copper ores are to be found in the mountainous areas of Eastern Turkey and Syria, in the Zagros Mountains on the western edge of the Persian Plateau, in Sinai, in the mountains of the Arabian Desert east of the Nile and, of course, on the island of Cyprus from which we derive its name. Of these ore deposits those in Eastern Turkey and Northern Syria lie fairly and squarely in the middle of the area in which we have seen early farming first developed and we need not be surprised, therefore, that it is precisely in this area that we find the first use of native copper for the making of small objects.

The urge to produce brightly coloured things found an outlet in another technological field. The earliest pottery we know seems to be purely utilitarian and its firing was not particularly carefully controlled. At the end of the firing process the pottery evidently lay buried or half-buried in a bed of smouldering wood and ash with the result that the surface is often multi-coloured with scores of black and grey across a body which is often brown or dull red in colour. But even amongst the earliest pottery we can detect an attempt on the part of its makers to do something about rectifying this rather dull appearance and we find for example that red ochre had been burnished into the surface of the pottery before it was fired so that it would ultimately appear a uniform red colour. The process was not always successful, and more often than not the surface was marred by black smeeches where the burning wood had fallen across the pottery and stayed there while it cooled. But at much the same time that early man was developing an interest in metallic copper he was also giving considerable thought to methods of producing pottery that would allow the colours

47 Painted pottery from Northern Syria, about 3500 B.C.

Red ochre was used to paint designs on pottery, its colour giving a strong contrast to a buff or white surface. The painting was clearly done with some form of brush, but it is questionable as to whether this was a brush specially made for the purpose or merely a feather dipped into the pigment.

to stand out unmarred. Essentially his problem was one of removing the pottery from the fire itself at least during the cooling stage, and initially this may have been achieved by removing the pottery from the fire on the end of a long stick and setting it down to cool in the open air – although later it was to lead to the development of a structure in which the pottery on the one hand and the fire on the other were separated. Furthermore, the potter was no longer content simply to have pots of red and his attention began to be drawn to different clays and the different colours they produced after firing. Potters now began to combine in the same vessel clays which fired white and clays which fired red under normal conditions, and we begin to find pots which are of a red body and which have been decorated on the surface with white clay or vice versa.

A few copper trinkets and a number of gaily painted pots, however, do not represent a technological advance, and their presence can have done little to alleviate the hard life of man. The importance of their appearance at this stage in man's history lies not in the end-products themselves but in the fact that through them we can see that the fire was becoming no longer a domestic hearth simply for cooking, lighting and keeping wild animals at bay: the fire was beginning to be appreciated as a place in which many raw materials could be greatly changed. If we are to understand why mankind moved on from this platform of technological development we must look elsewhere for the answer.

As we have already seen, the earliest farming communities that we know of tend to lie in upland areas – in regions in fact in which the soil is relatively light and easily worked. But these same areas have a disadvantage from the farmer's point of view in that the land may rapidly become exhausted. Thus, after a clearing had been made in the woods to establish a field, the first few crops were probably of good quality, but in time the fertility of the soil fell and new clearings must have been made. This in itself restricted the possible size of any single community if the fields themselves were not to become far too distant from the village to be of practical value, apart from which hunting remained an important aspect of the economy and much of the foodstuffs still had to come from wild animals, while the virgin forest in which the game lived became ever more distant from the habitations. In turn, this would have meant that adjacent communities could only thrive at considerable distances from one another. It would be totally erroneous to suggest there was any shortage of land – that there certainly was not – but farming in the upland areas must have given rise to somewhat isolated settlements.

It would be wrong, too, to suppose that early farming people were a static population rooted to a particular piece of soil as was the case with so many later peasant communities. Within the few centuries immediately following the year 5000 B.C. we find them spreading into Central Europe – eastward into the Indus Valley and southward down into the Nile Valley, and it is in the valleys of these great rivers that are to be found the greatest technological changes. The rivers themselves not only acted as roadways connecting each community with the next but they also watered the crops and, when in spate, brought down in their waters the essential minerals required to maintain the fertility of the land. Briefly, in these river basins it now became possible to crop the same field year in and year out without having to allow it to lie fallow and without having to cut new fields into the forest. Indeed, within a few centuries mankind had in all probability completely transformed the appearance of these valleys.

In theory, with fields that could be cropped perennially it now became possible for the size of individual communities to grow. The village that until now had contained at the most a few hundred souls could now contain a few thousand. But in the ability to expand there lay a danger. Were all the land available for agriculture in the immediate vicinity of a community given over to that purpose the livestock would have to be fed in the woodland beyond, and between them cattle and sheep can, if contained within too small an area of woodland, reduce it to a waste in a very short period of time. Cattle are essentially branch feeders, eating the green leaves of trees and occasionally the bark, while sheep are ground feeders, nibbling at the newly growing shoots and undergrowth. Thus, cattle denude the standing forest and sheep prevent regeneration. Once forest had been destroyed by the grazing of herds and flocks there was an ever-present danger that the land would be given over to agriculture. Thus, two small villages which were initially at a comfortable distance apart might, if the population grew too greatly, ultimately find themselves in dispute, not so much over the land held for agriculture as for the residual forest used for grazing. Alternatively, all the land between separate villages might have been used for agriculture while the herdsman had to look further and further afield, to such an extent that the agriculturalist and herdsman became essentially distinct communities.

This state of affairs was not reached, of course, in a matter of a few years and not necessarily in a matter of a few centuries, but it was ultimately to result in an extremely dense population in these river valleys.

Even from the outset, however, the same system of agriculture could not be followed exactly in both the Nile and Euphrates Valleys, for the two rivers flood at different times of the year. The Nile floods in the early spring and the date at which it does so lies within very narrow limits year after year, a matter of a few days. Thus, the river can be allowed to flood and once the waters have receded crops can be planted, and it remains then only to ensure that the growing crops get enough water. The Tigris and Euphrates, on the other hand, are in spate in the early summer and once their waters have receded it is too late to start planting. Sowing, therefore, has to take place before the rivers flood, with the result that there is a very real danger that the whole crop may be inundated and completely lost. If agriculture in the Mesopotamian Valley, therefore, were to be a success the fields had to be defended with dykes against flooding. Early written sources show us that flooding was an ever-present menace to the farmers of the Mesopotamian Valley, while some authorities believe that the story of Noah and the Great Flood was a legend that the Hebrews acquired from the Mesopotamians. But unfortunately we know very little about early agriculture in the Mesopotamian Valley, for later work has almost totally obscured what dykes and drainage canals were built by these early farmers, and it is not until we come to written sources that we begin to appreciate fully the amount of irrigation work that had to be carried out by these people if agriculture were to succeed.

While in Mesopotamia dykes had to be built to prevent flooding, in Egypt canals had to be dug in order to water the land after the floods had receded. Again it is not until we come to written sources telling of the first legendary king, who is reputed to have organized an enormous system of canals, that we get any first-hand evidence of the methods of irrigation in Egypt.

Grain crops were not the only plants grown by these people in the river valleys. The occasional finds from tombs and villages of the remains of food show us that grapes, the olive, figs and dates were all being cultivated. Mankind was hence becoming not only a farmer but a gardener as well. Thus, with fields that could be cropped repeatedly year after year, more long-standing policies such as the establishment of orchards, which could be included in the general scheme of irrigation, became feasible.

Before the year 4000 B.C. mankind had shown an interest in two metals other than copper – gold and silver. Unlike native copper, gold is not found in association with ores but as veins usually in quartz rocks. When gold-bearing rocks become weathered and eroded the metal with the other mineral detritus is carried away by streams. The larger fragments of gold, the metal being relatively a very heavy material, are seldom carried far by water and so become deposited in the alluvial gravels of the upper reaches of the rivers. These so-called placer deposits often contain nuggets of gold the size of a pea or even larger, while further downstream only the finer particles will be found distributed as dust through the river sands. Theoretically, then, gold in the form of placer deposits had been available to man since his remote origins, and it may even seem surprising that it had not earlier attracted his attention. On the other hand, while the mountainous upper reaches of rivers are unlikely to have been particularly attractive to the hunters of big animals, the foothills surrounding the highland plateaux on which early farming communities were so often situated may have been sporadically visited from time to time by shepherds. Vein gold – that is to say the metal embedded in quartz rocks – was clearly not being worked at this early period for it required a great deal of heavy equipment to remove the metal from its matrix, and one can only assume that placer deposits were being panned for the metal. Panning was a fairly simple process in which some of the sand or gravel was placed in an open dish with water and swirled round so that the lighter detritus slopped over the edge leaving the metal, if any, in the bottom of the pan.

Gold, unlike copper, does not harden when hammered and it is possible to beat the metal out into relatively thin sheeting without the need to reheat it from time to time to prevent it cracking. Furthermore, unlike copper, gold can be made to join simply by hammering two pieces together, when a perfectly good weld will form. Nevertheless, early objects of gold are invariably very small and it seems probable that this is a reflection of the rarity of the metal rather than of the fact that this property had not yet been discovered.

Native silver occurs far less commonly than gold, although often the two metals are found together as a natural alloy, electrum, in which the proportions of the two metals vary very considerably from area to area. Today these natural alloys would be classified as gold and the two metals would be separated, but to early man they were worked as they came, with the result that without chemical analyses it is often difficult to know whether small objects from the remote past were made of pure native silver or merely an electrum containing a low quantity of gold, although a few pure silver objects of this early date are known and it is quite clear that early man was on the look-out for this metal too.

48

49

One other brightly coloured rock was evidently much prized by these early farming people – the blue mineral lapis lazuli with its vivid ultramarine colour. Its natural occurrence is extremely rare, and its source is still far from clear, but we know that it was in short supply and that at some time before the year 4000 attempts were made to produce a synthetic lapis lazuli. This synthetic material, usually referred to as Egyptian faience because such enormous quantities of it were

48 A tablet of 'faience' from Egypt, before 2000 B.C.

Beads and small ornaments were made in imitation of the blue stone, lapis lazuli, by heating shaped pieces of soapstone in the presence of ores or copper which in effect covered the objects with a turquoise glaze. Later the soapstone core was replaced by a synthetic material (possibly mankind's first) made by heating quartz-sand and soda until the sand particles fused. This material is generally known as 'Egyptian faience'. The work was probably carried out in fires similar to those used for softening metallic copper.

49 Faience-making in Persia today

'Egyptian faience' is still being made in remote parts of Persia today. Although the process has obviously changed in details over the ages, it is probably basically the same as that used in 4000 B.C. The objects to be glazed are heated in a lidded clay vessel packed in a mixture of wood ashes, lime and green copper ore.

later produced in Egypt, can be looked upon as man's first real move into the world of synthesizing the material he required. In fact the blue colour of lapis lazuli is due to a highly complex molecular arrangement and the material was not truly synthesized until the nineteenth century. Early attempts at reproducing it were based upon the blue colour that can be derived from copper when it is incorporated into a glass. The early examples of Egyptian faience appear to have been made by powdering the surface of talc-stone with one of the ores of copper – probably azurite or malachite – and then heating so that the whole surface of the object became a blue-coloured glass. In order to accomplish this a number of conditions had to be provided. First of all the material could not be made in an open fire: the work had to be carried out in a crucible or some other lidded vessel which would keep the smoke and ash away from the object being made. Secondly, a much higher temperature than is normally found in the domestic fire would be required. In short, the fire would have to be fanned or blown, and while we have no idea precisely how this was done, at a very much later date we find metal-workers in Egypt, for example, attaining these high temperatures by blowing on the fire through hollow tubes and it seems highly likely that even at this early date a reed was used to force air into the fire and so raise the temperature sufficiently to allow the work to be carried out.

Egyptian faience is still today being made in some of the less accessible parts of the Near East, as for example in Persia. The process is seen in its simplest form where the articles to be glazed are packed into a lidded

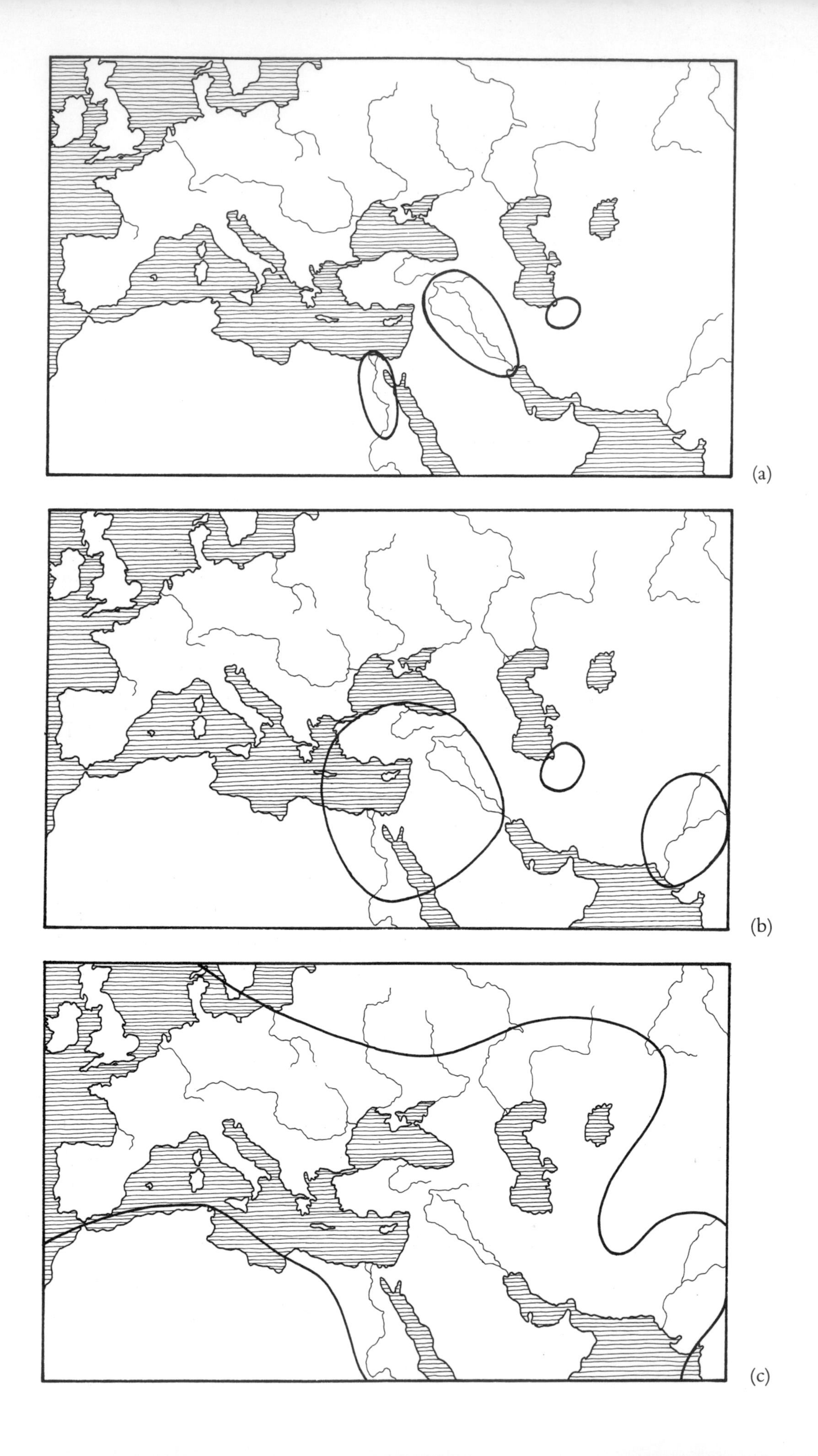
(a)
(b)
(c)

50 The spread of 'Egyptian faience': (a) up to 3000 B.C.; (b) 3000–2000 B.C.; (c) 2000–1500 B.C.

The earliest objects of 'Egyptian faience' were made about 4500 B.C. in the Tigris–Euphrates Valley. By 3000 B.C. they are found in the Lower Nile Valley and in Northern Iran, where they are probably imports. A thousand years later they are found more generally throughout the Near East and as far afield as the Indus Valley. By about 1500 B.C. beads of 'Egyptian faience' occur all over Europe where they were undoubtedly acquired by trade with the Near East. By this time Egypt had a thriving 'faience' industry of its own.

clay vessel surrounded by a mixture of lime and wood-ash to which has been added a small quantity of copper carbonate. The vessel is then heated for a day to a temperature of 950°C – only just below the melting-point of copper – and on cooling the objects are removed covered with a somewhat uneven blue glaze.

The importance in the history of technology of the appearance of Egyptian faience does not lie in the immediate effect that the material had upon those living at that time – in fact it probably had no worthwhile effect at all – but it did provide a set of conditions that may ultimately have led to the smelting of copper from its ores.

At about 4000 B.C. – in fact at much the same time that we find Egyptian faience appearing in Mesopotamia, long before it was made in Egypt – we find also the appearance of true pottery kilns. As has already been pointed out, for polychrome pottery to be a success it was essential to separate the vessels from the fuel during firing. No doubt a number of different arrangements were tried, and indeed some of the early kilns show very considerable variation. But shortly after 4000 B.C. we find throughout Mesopotamia a quite standard form of kiln arrangement. The fire was lit in a hearth underneath the kiln chamber and the vessels were separated from the fire by a floor which was perforated in a number of places. Unfortunately all the kilns so far excavated have been severely damaged and the upper parts are generally missing, but we would probably be safe in assuming that they resembled the kilns that we know of some five hundred years later, which appear to have had a dome-shaped cover to the chamber with a vent at the top. These structures were built very largely of clay, usually with an outer wall of either stone or mud-brick. As one might anticipate, as a result of firing pottery in these kilns the structures themselves also became baked hard and are often far better preserved than the buildings around them made only of unfired brick. Apart from the obvious

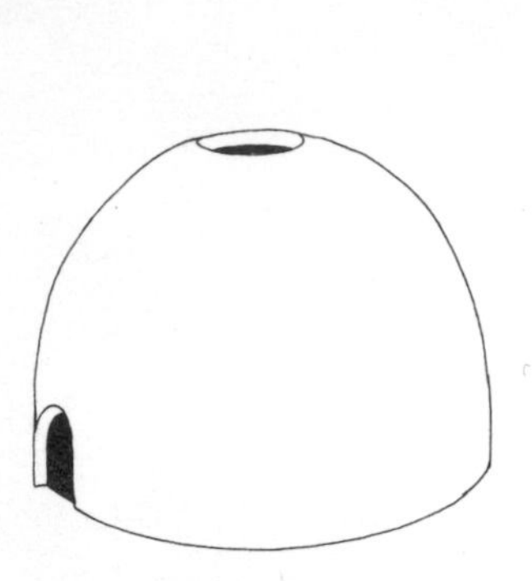

51 Reconstruction of a Mesopotamian kiln of about 3500 B.C., based on a number of excavated examples

The decorative effect of painted pottery would have been spoilt had the wares been fired in an open fire, for ashes and pieces of wood lodged on the surfaces would have caused smeeches. The use of kilns, in which the vessels were placed on a perforated clay floor above the fire, overcame this difficulty. To judge by the small number of kilns so far excavated, in Mesopotamia they were usually low, domed structures with a vent at the top.

52 Reconstruction of an Egyptian kiln of about 3000 B.C., based on a number of tomb paintings

By contrast early Egyptian kilns, which we know of only from tomb paintings, appear to have been taller, chimney-like structures, open at the top. In some cases a platform was built to one side to allow the potter to stack his wares through the top opening. How the pots were set is a matter of conjecture: they may have been piled in haphazardly or projecting platforms may have been built on which they could stand. The opening at the top appears to have been partially closed with a capping of mud or stone.

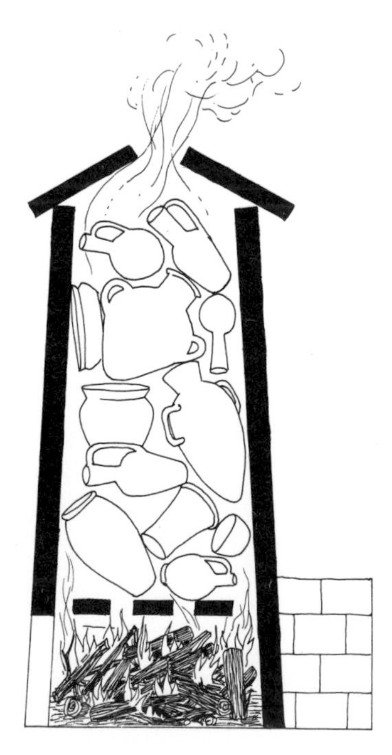

53 Egyptian potters at work: a tomb painting of about 2000 B.C.

54 Reconstruction of a potter's turn-table of about 3000 B.C. based on surviving fragments from Mesopotamia and on Egyptian tomb paintings

In the scene from an Egyptian tomb-painting one of the tall kilns can be seen on the right being loaded with pottery. On the left is a second kiln being fired. Two men can be seen treading clay to make it homogeneous, while the potter can be seen squatting at a low turn-table on which he is making his wares. Next to the turn-table is a pile of prepared clay, and beyond stands an assistant ready to remove the vessels as they are made. The turn-table was probably made of wood or clay, and was pivoted into a socketed stone at the base. A mound of clay was piled on to the turn-table, and from the top of the mound the potter shaped his wares, each vessel being cut off and handed to the assistant when finished.

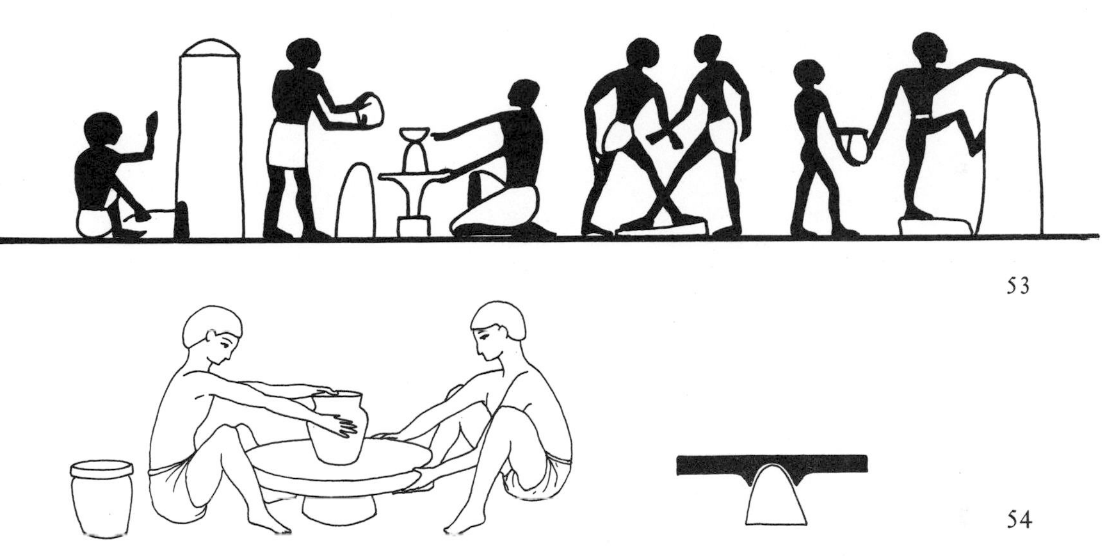

advantages of being able to separate fire and vessels, the kiln was also presumably less extravagant on fuel, and furthermore allowed a larger scale of pottery production.

Possibly to meet the growing demand we now find a new tool introduced in the manufacture of pottery. Up to this time all pottery had been made by a number of simple means – by moulding slabs of clay into pre-formed moulds, by building up the walls of the vessel with successive rings of clay or by some other rather tedious method which required a considerable amount of hand modelling. The new device that was introduced shortly before 3500 B.C. has sometimes erroneously been called a potter's wheel, but in fact it bore little relationship to what we know as the wheel today. More correctly it should be called a turn-table. Its function was quite simple. Essentially it was a flat disc that could be made to rotate around a central pivot. A ball of

55 A modern potter, Sind

This system of shaping pottery can still be seen in some parts of the world today, as, for example, in Sind, although here the wheel is larger and set lower than that in the picture from Egypt.

clay was placed in the centre and while the disc was made to rotate with one hand the other was used to shape the clay into the form of the required vessel. Unlike the modern potter's wheel, this pivoted disc obviously could not be made to rotate continuously, but nevertheless it greatly speeded up the method of production, and furthermore introduced a greater uniformity than before into the style of pottery being made within any single workshop. The turn-table remained the chief means of pottery production for many hundreds of years after this time, and in fact we shall discover that it is not until perhaps 700 B.C. or even later that it was superseded by a true wheel that would rotate continuously.

The turn-table and the kiln between them set new demands upon the potter so far as his clay was concerned, for the body of his pottery could no longer be as rough and as coarse as before. Large coarse particles incorporated into the body material would have tended to have got caught in the potter's hands as the clay rotated on the turn-table, and once dislodged would have cut through the walls of the pot he was making. The material had thus to be of far finer texture, and it had, too, to be more uniform, for clay that is inadequately mixed can only be used as long as there are large coarse particles in it to prevent excessive shrinkage and to make it porous during firing. But a clay containing fine particles that is poorly mixed will shrink unevenly as it dries, and furthermore lose water at different rates during firing with the result that steam will build up in the walls of the pottery which may cause the pot to burst, setting off a chain reaction in the kiln. To meet these new conditions the potters developed the practice of levigating their clay – that is to say they mixed the raw clay thoroughly with water and allowed it to settle, tapping off only the finer clay fraction from above and leaving unused the coarser lumps that had settled out at the bottom of the levigation pit. Thus, by 3500 B.C. the making of pottery was beginning to become a highly complex industry requiring a considerable amount of know-how.

One small device appears at about 4000 B.C. which is of considerable interest in the history of technology and ultimately was to lead to one of the most important inventions of all time. This was the seal. Early seals were small circular discs of fired clay or stone into which was cut an impression of the mark required – usually some geometric pattern – while on the back of the disc was a small perforated lug used as a handle when making the impression. The device was clearly an owner's

56 The back and face of a stamp seal from Mesopotamia, about 3000 B.C.

Small seals, usually of stone, were used to stamp the owner's mark into wet clay as, for example, when stoppering jars and bottles.

mark much like the seal in medieval and later Europe. It was, one imagines, usually impressed into wet clay as, for example, when sealing jars or bottles. While ultimately the seal was to lead to a form of writing, for the moment it is important to notice that the idea of producing a design in negative was fully appreciated by this time.

The stage was now set for two most important discoveries – the fact that metallic copper could be reduced from its ores, and that once reduced molten copper could be poured into moulds. A great deal has been written about these two discoveries and most of what has been said has been speculation. On the whole it would seem unlikely that we shall ever know precisely when and where these steps were taken. Certainly, copper ores could easily have been reduced to the metal during the process of making Egyptian faience or even during the manufacture of pottery – although the former seems more likely. The process requires of the metal-worker some understanding but not a great deal of equipment. The ore was mixed with the fuel – ideally charcoal, although any very dry wood would have sufficed – and placed in a shallow pit in which a small fire had already been kindled.

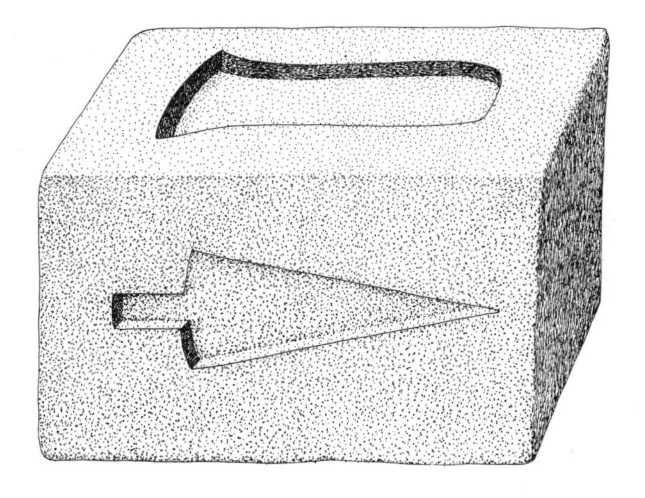

57 Reconstruction of a simple mould for casting spear- and axe-heads

58 A copper axe-head that would have been cast in such a mould, from Palestine, about 3000 B.C.

With the discovery that copper would become fluid at sufficiently high temperatures it became a practicable possibility to shape the metal by pouring it into moulds. The concept of casting already existed as in the making of mud-bricks (Figure 26) and in the seals above. At first moulds used for casting copper objects were simple depressions cut into the surface of a suitable stone. Objects cast in such moulds still needed a great deal of hammering and polishing to bring them to the required shape.

59 Reconstruction of a two-piece mould used for casting more complex shapes

60 A tanged knife that would have been cast in such a mould, from Palestine, about 3000 B.C.

By using two opposing mould pieces it was found that a great deal of the final shaping by hammering could be avoided. The moulds were made either of stone or of fired clay, the two pieces being held in the correct position by means of short dowels.

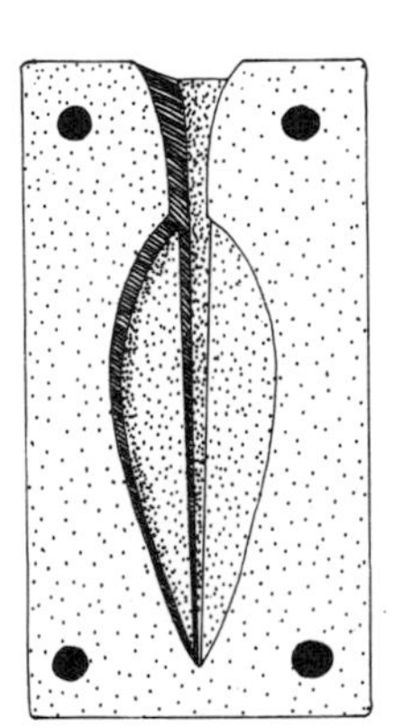

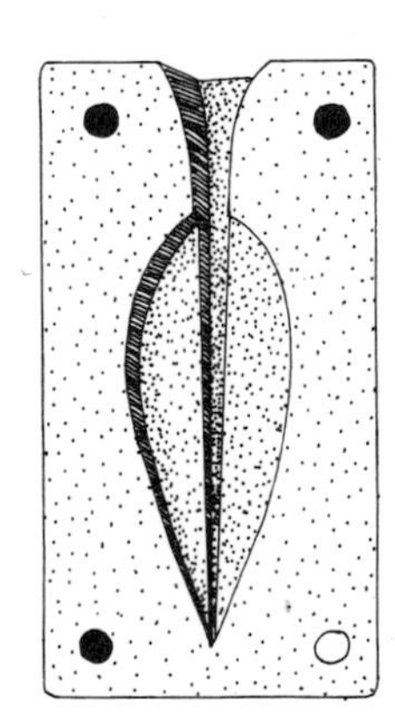

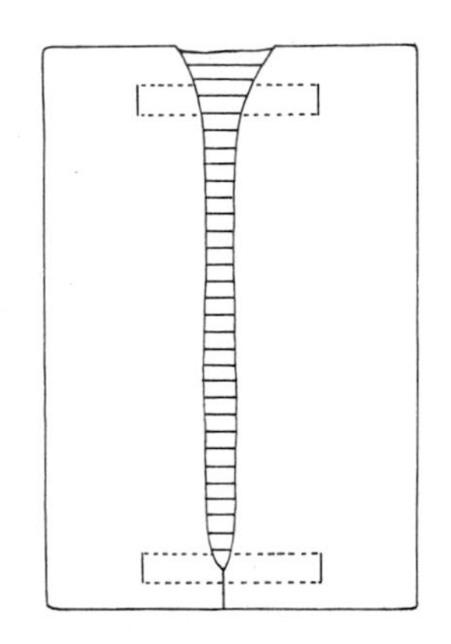

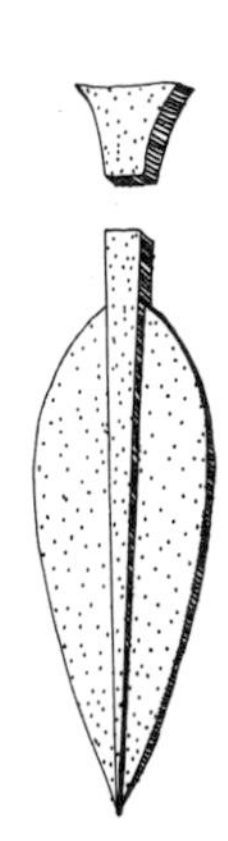

The fire then had to be brought to a white heat using the blow-tubes already mentioned. The high temperature had to be maintained for a considerable period – probably the best part of a day – and then the whole fire was allowed to die down. When cool, the molten copper would have been found to have settled to the bottom of the pit while above it would have lain a layer of glassy slag that would have been chipped away and discarded. The metal as it stood would have been full of blowholes and very unpleasant to work, although with a great deal of hammering and annealing it could have been shaped into objects of various forms. On the other hand, if the ingot of blistered copper were broken up, placed in a crucible and reheated in a small furnace of the type just described it would have become fluid and could have been poured out, when the worst of the blistering would have disappeared. At this time the mould, admittedly used for the making of bricks, was very nearly two thousand years old and stamp seals, as we have just seen, were already in use. It is hardly surprising, therefore, that at the same time that we find early man reducing metallic copper from its ores we also find him designing moulds specifically for the casting of that metal. At first these were very simple devices. The approximate shape of the object to be cast was cut as a negative into a suitable piece of stone such as talc-stone or a fine-grained sandstone, indeed any rock that would stand the high temperatures involved, and

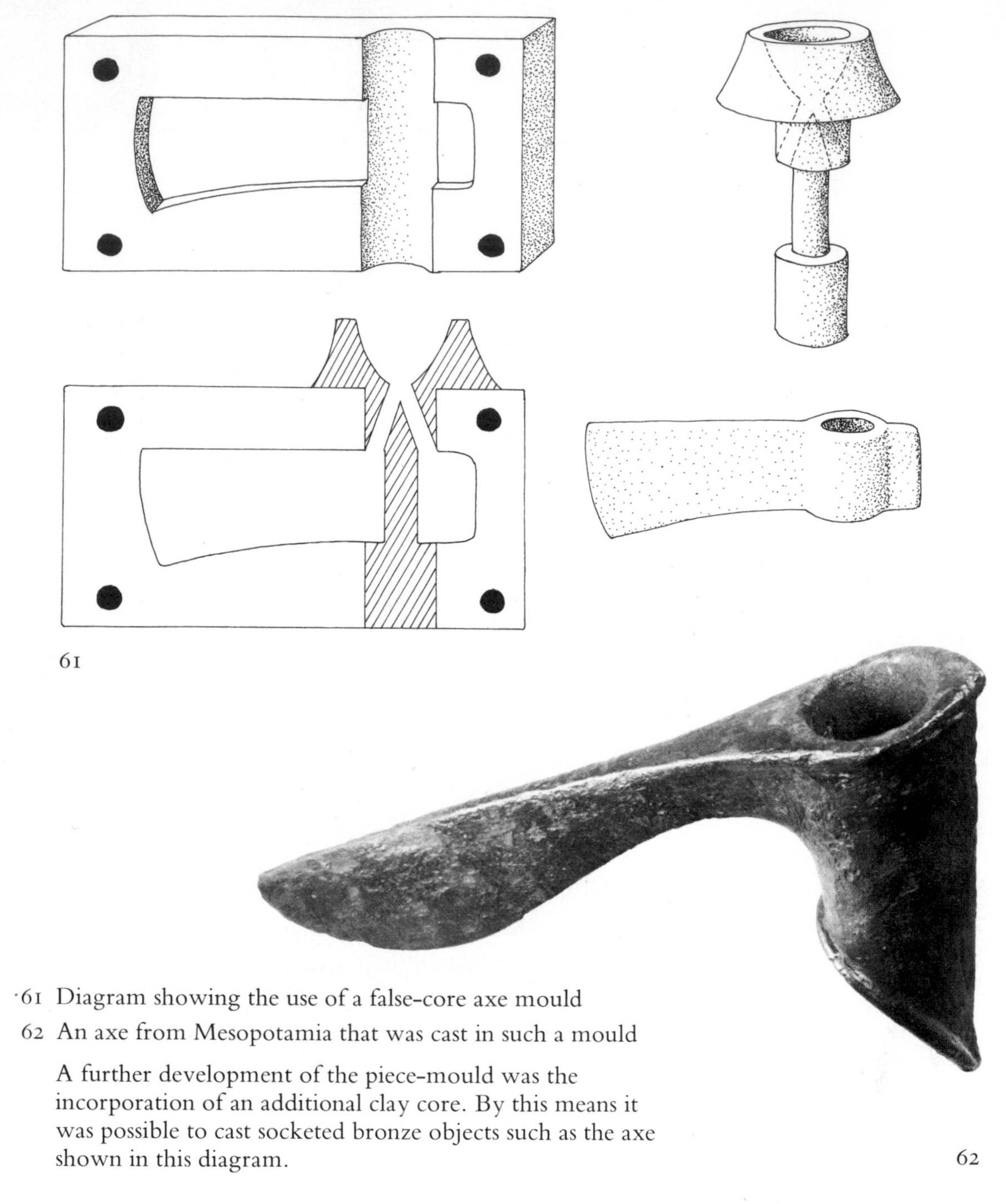
61

62

·61 Diagram showing the use of a false-core axe mould

62 An axe from Mesopotamia that was cast in such a mould

A further development of the piece-mould was the incorporation of an additional clay core. By this means it was possible to cast socketed bronze objects such as the axe shown in this diagram.

63 Egyptian metal-workers from a tomb relief of about 2500 B.C.

64 Diagram showing how the crucible was probably held

With the discovery that copper could be extracted from its ores, the metal became more plentiful. This scene from an Egyptian tomb shows copper being melted on the left, and on the right being poured from a crucible into the mould. In this case it almost appears that the red-hot crucible is being held directly in the hands of the founder. Probably it was grasped between two stones as shown in the accompanying drawing.

into this depression was poured the molten metal. Initially, no attempt was made to produce in detail the shape of the object required. Instead a blank was formed which later would be hammered and annealed until the required shape was achieved. But within a span of a very few centuries new methods of mould designing were being evolved. The smith had learned to cut moulds of two opposing pieces bearing the larger part of the required design of the finished object. The two pieces were held firmly together and the metal would be poured into the space between them in such a way that when cool an almost perfect form of the object had been produced. Furthermore, the two-piece moulds were often no longer made of stone but of fired clay. In such cases the smith thus began by making a pattern of the object he required to cast, often of wood, and around this pattern was built first one then the other piece of the clay mould. The pattern was then removed, the mould pieces were fired like pottery and reassembled, and into the space once occupied by the pattern was poured the molten metal.

While the earlier form of mould – the simple depression cut in a piece of stone – allowed only limited shapes to be produced, since the upper surface in the mould always had to be flat, the later piece-moulds allowed a far more elaborate series of objects to be manufactured. With increasing supplies of copper from the process of smelting, and with improved methods of casting, copper objects now becamc quite common although the metal was probably still only available to very

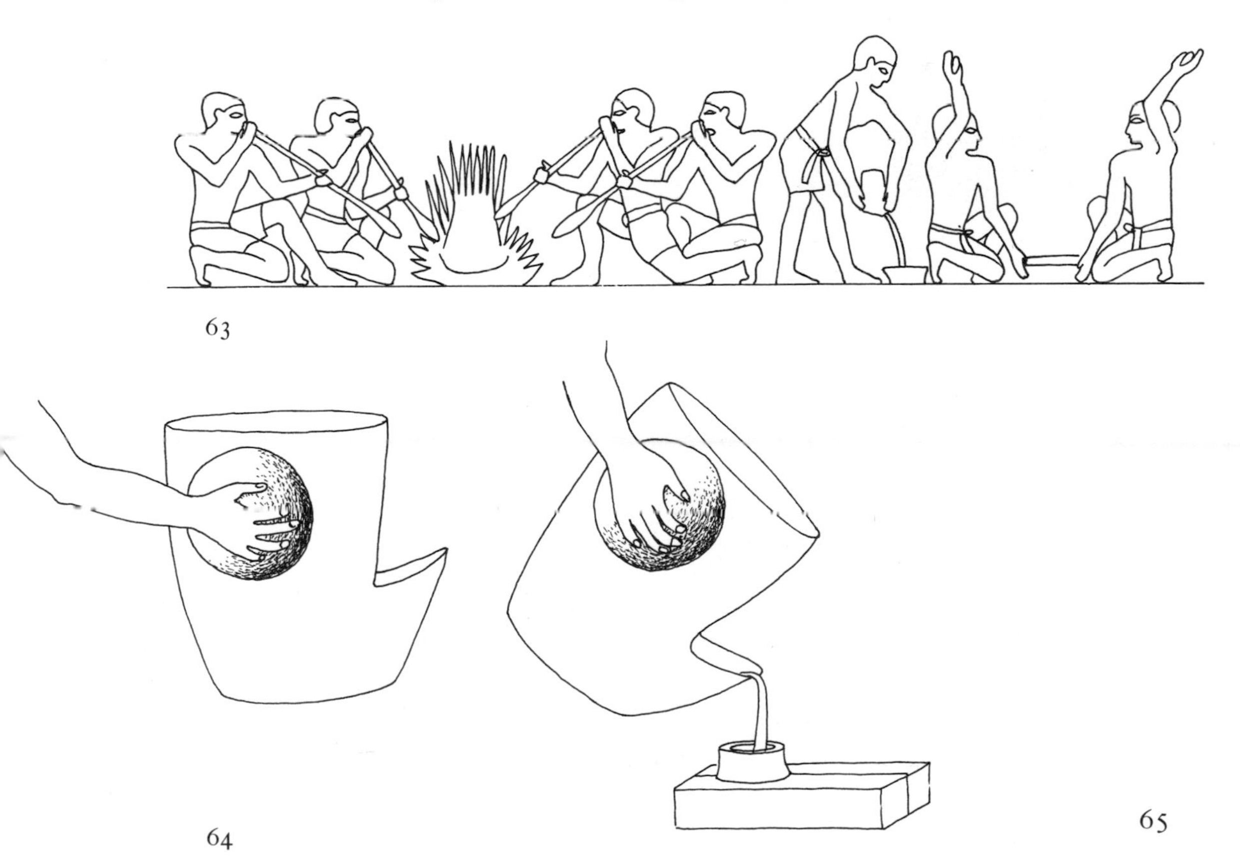

63

64

65

65 A helmet of sheet gold and a cow of copper: fine examples of Mesopotamian metal-working of about 2500 B.C.

Despite improved casting methods, the hammer still remained the major tool for shaping metals. This helmet, for example, was made entirely of hammering, while the decoration was added by chasing the surface with punches.

wealthy people, and hence apart from ornaments and trinkets we find the use of copper limited almost entirely to the manufacture of weapons. It was still too rare to be used extensively for the manufacture of tools.

With the extensive use of kilns for firing pottery and furnaces for making metal objects, it is hardly surprising that by 3500 B.C. in Mesopotamia, at least, experiments were made with the firing of bricks. Hitherto all the building in this area and in Egypt had been carried out in sun-dried mud-brick, and although this allowed the construction of fairly large buildings it had the very great disadvantage that in the winter frost and rain caused the surface to disintegrate. Losses on the surface of walls due to weathering could of course be plastered over with more mud from time to time, but in due course the building was bound to become both unsightly and unserviceable. In fact at this point in time villages, particularly in the lower valley of Mesopotamia, can no longer be looked upon as simple settlements serving as markets but must be seen as embryonic cities with an administrative body that insisted on being housed in a suitable style. Thus every small city had its own central temple and it was, apparently, the temple servants who became the administrators. A dilapidated mud-brick temple in the middle of the city would of course have been a contradiction in terms, and it was to these often quite substantial buildings that the new technique of using fired brick was applied. Early forms of sun-dried mud-brick had normally been on the rather large side – as a rule measuring about eight inches by sixteen inches by two and a half inches. Such large bricks clearly proved too difficult to fire, and we now find a far smaller brick being used, very similar in proportion to our own modern bricks. Furthermore, various shapes of bricks were also being made, as for example slightly curved forms for building pillars. But it would be quite wrong to imagine that the whole temple was built of fired bricks, for often they were used where one would expect a great deal of wear and tear. A second quite different solution was found in order to prevent the excessive weathering of sun-dried mud-brick walls. Into the relatively soft surface of the wall was hammered a number of cones,

66

67

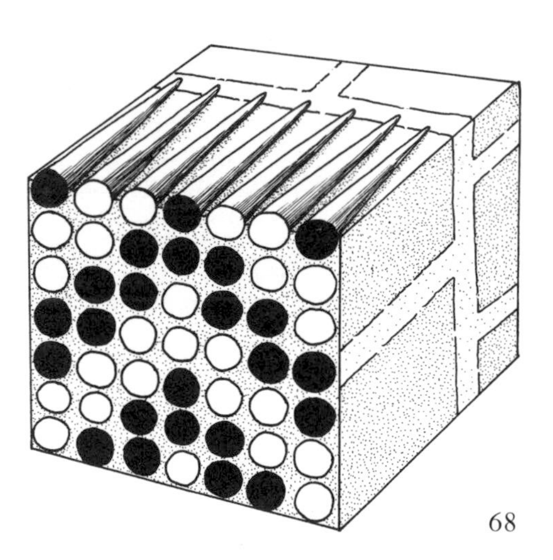

68

66 A brick vault from Ur, Mesopotamia, about 2500 B.C.

Fired bricks, of smaller dimensions than the sun-dried mud-brick, were used to a limited degree especially to provide decoration or to give added strength to the building. The use of fired brick made possible the development of true arches as seen in the vault of this tomb from Ur. For reasons of economy however, sun-dried brick continued to be used for the major part of all buildings.

67 Cone mosaic from Mesopotamia, about 2500 B.C.

68 Diagram showing how the cones were hammered into the soft mud-brick wall

Cones of fired-clay or of stone were hammered into mud-brick walls to produce decorative effects. Walls treated in this way were naturally far less prone to wear and tear than those of unfired brick.

some of which were made of fired clay, others of differently coloured stones. These cones were usually set into the wall to form a pattern, their exposed heads making a façade that gave the appearance of being a mosaic.

The administrators of the city's affairs were, however, faced with a far greater problem than that of simply keeping the temple in good order, for each city now controlled a sizeable acreage of land and many cities had quite a number of dependent villages. To administer such a community properly it was found necessary to be able to keep records of such things as the taxes to be paid and the services to be provided by individual citizens or smaller dependent communities. We have already seen that the stamp seal pressed into clay was being used as a means of identification by 4000 B.C. and now, five hundred years later, we find the palace servants adapting this idea as a means of keeping records. In brief, we find the beginnings of a system of writing in which signs were made on slabs of wet clay which were dried and stored as a record. The symbols used were not made with a stamp, however, but were drawn free-hand. Initially the symbols used were pictograms. That is to say, if one wished to record the payment of a sheep then a picture of a sheep was drawn. But if confusion, of the sort so adequately described by Kipling in his story of the 'First Letter', were to be avoided, then agreement had to be reached, for example, on how a sheep was to be drawn so that it could not be confused with any other animal. Indeed, in the earliest period of recording, we find several different symbols used to denote sheep of different kinds – lambs, yearlings, ewes, rams and so on. Later a single symbol was used to denote all sheep, and a series of other signs was appended to indicate what type of sheep was meant. In the course of time these pictograms became greatly abbreviated and simplified to such an extent that without a record of the various stages of abbreviation it is often difficult to see any resemblance between the symbol and the original pictogram.

Drawing easily decipherable symbols with a pointed object on the surface of a piece of wet clay is not an easy matter, and in Mesopotamia the ultimate solution was found not in the making of free-hand drawings, but by impressing into the surface a small stylus cut from a length of reed which was triangular in section, and which thus made small wedge-shaped impressions in the clay. The free-hand symbols were further modified to this system of writing so that every symbol was now composed of a series of differently positioned wedge-shaped impressions. The end-product seems even more remote from the original pictogram than before. In the subsequent centuries this system of writing, referred to as cuneiform, spread throughout a larger part of the Near East, although it was never greatly used in Egypt where a different writing material, as we shall see later, led to a rather different system of recording.

From our point of view the records from the temples of the various cities of Mesopotamia come as a salutary reminder of what we have overlooked in this account of early technology. On the clay tablets are depicted a number of objects, and even pieces of equipment, of which there is little if any record in the way of material remains. Two such pieces of equipment are the plough and the cart, both of course drawn by beasts of burden. Unfortunately we remain far from sure precisely what draught animals were first used or, indeed, how they were harnessed. Horses, which are naturally steppe dwellers, were certainly not domesticated in Mesopotamia at this period and it seems likely, therefore, that the earliest beasts of burden were the ox and the wild ass, or onager. Of these two it would appear probable that the ox was the first to be made to draw man's vehicles. This is suggested only because the earliest illustrations of ploughing in Egypt show oxen harnessed to a plough not by a shoulder-yoke but by one attached in front of the animal's horns, and the shoulder-yoke, and by inference the use of onagers for pulling vehicles, would seem, therefore, to be a later development. Inefficient, and even cruel, as the harnessing of oxen by their horns may seem to us today it could well have been used as a means of shifting heavy weights long before the introduction of either the plough or the wheeled vehicle, and the earliest pictogram inscriptions from Mesopotamia show yet a third vehicle, the sledge. Indeed, the pictogram used to describe the wagon is simply a sledge

69 The earliest form of writing on clay tablets from Mesopotamia, about 3000 B.C.

70 A cuneiform tablet from Mesopotamia of about 800 B.C.

With the growth of cities in Mesopotamia the administrators found it necessary to keep records. At first they were made by sketching commonly understood symbols (ideograms) on the surface of tablets of damp clay. These, when dry, were stored in cellars so that many have survived.

In time the practice of sketching the ideograms gave place to a method in which each figure was built up by impressing a reed of triangular section into the surface of the clay to give, ultimately, the form of writing known as cuneiform. The ideograms bear no immediate resemblance to the sketches from which they were derived.

69

70

71

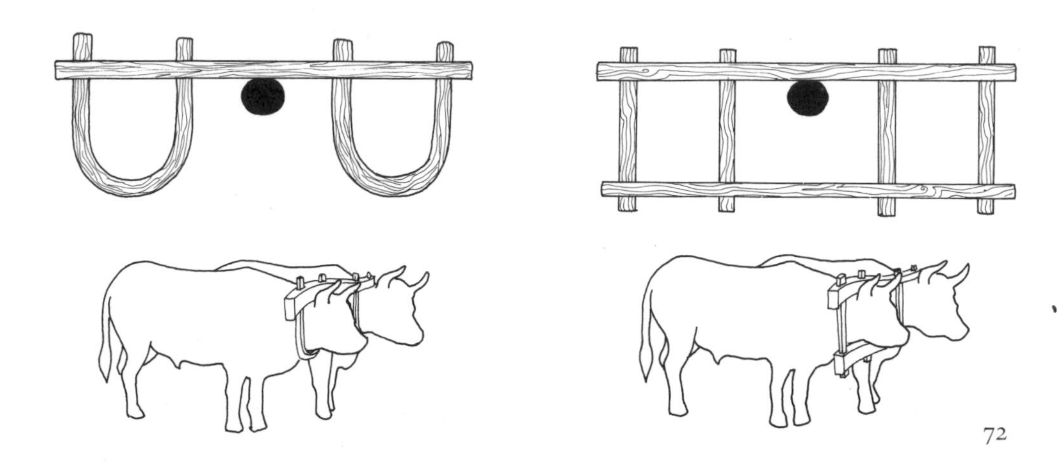

72

with the addition of two pairs of wheels. Sledges of various types continued to be used in the Near East for moving very heavy weights for a very long time thereafter, as we shall see later.

In the five hundred years that follow the year 3500 B.C. we know of a limited number of two-wheeled carts and four-wheeled wagons as actual vehicles, from pictures and as small pottery models. The wheels of these vehicles were made of solid wood, but even in the earliest forms in which we know them they were never made of a single piece of timber but always of three planks joined by two cross-struts. The middle plank, which was initially larger than the other two, had been shaved away to form a heavy central hub, and to this two part-circular planks were attached, one on each side, by means of cross-struts. The fact that this pattern of wheel was copied in other parts of the world to which it spread, although the construction of the carts and wagons themselves differed widely, suggests that the idea of this type of wheel construction diffused from a central point, namely Mesopotamia. It also seems that the axles were not attached permanently to the vehicles but were held in place by straps or some other device so that they could be removed readily, a feature which suggests that it was not an uncommon event to have to take one's cart to pieces when the going got too rough, and to reassemble it once the obstacle have been overcome

71 Cattle yoked by their horns, from an Egyptian tomb painting of about 2000 B.C.

Oxen were probably the first animals used for pulling sledges and ploughs, the yoke being lashed to the animals' horns rather than placed over the shoulders. How the yoke was tied to the horns of a pair of oxen can be seen clearly in this painting from an Egyptian tomb which, although late, probably depicts the earliest form of yoke.

72 Two types of shoulder-yoke for oxen from antiquity

The yoke placed over the animals' shoulders was, however, far more efficient since it allowed the oxen to exert a greater pull. The ox-yoke, once placed over the animals' shoulders required no further harness to keep it in place. Of the two types of shoulder-yoke, that on the right is clearly an element of the pictogram of ploughs in Mesopotamia (Figure 120). The other type appears to have been used commonly in Egypt.

73 The remains of solid wooden cart-wheels found during excavations at Ur, about 2500 B.C.

It is still far from certain where the first wheeled vehicles were developed. The first examples that are known come from Mesopotamia. Here the wheels are made of solid wood, but they are not made of one single piece. Instead they were built up from three sections, so it would appear that there may well have been earlier wheels, made of a single piece of wood, about which as yet nothing is known.

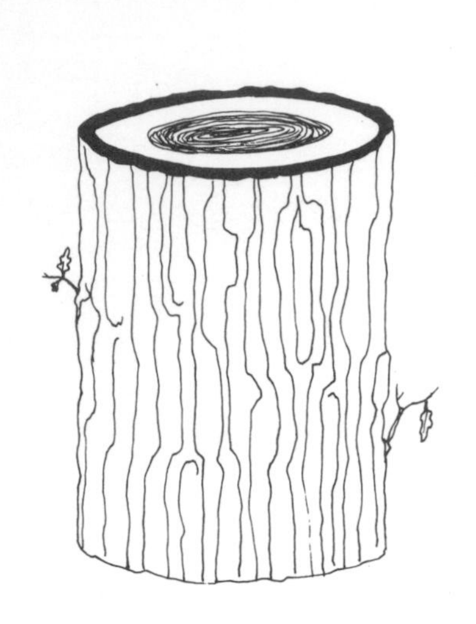
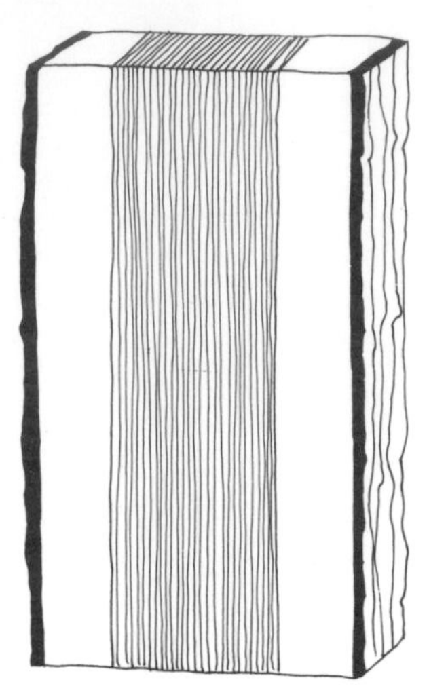
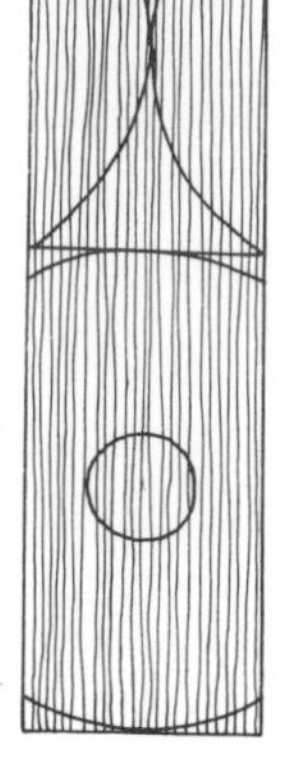
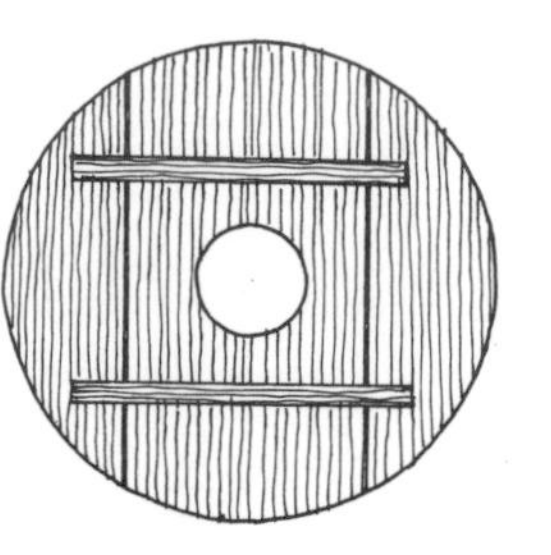

74

75

76

74 Diagram showing the process of making a three-piece wheel

75 An Irish cart, County Tipperary

Wheels made from three sections of planking are still made in many parts of the world today. Usually a plank is cut from the centre of a log, and the outer, soft sap-wood, which would wear rapidly, is removed. The plank is then halved: one half forms the middle plank of the wheel; the other half is further divided to form the remaining two sections

76 Limestone relief from Mesopotamia of about 3000 B.C. showing a cart drawn by onagers. The wheels were made in three sections

In Mesopotamia wild asses, onagers, were used to pull carts as well as oxen. The yoke, did not fit comfortably over the onagers' shoulders and had to be held down by a neck-strap which probably had a tendency to choke the animals whenever the yoke rode up. To guide the onagers copper rings were put through their noses, to which reins were attached. Although in Mesopotamia carts with two and four wheels were in use, no such vehicles appear to have been employed in Egypt where, one supposes, boats and rafts on the Nile were the main means of transport.

It follows that early carts and wagons were not used for long-range transport but rather for shifting heavy loads over short distances, either from outlying villages to the central city or perhaps even from one adjacent city to another.

Even so, it must be admitted that there may be a very considerable period in the early development of the wheel about which we know nothing. On the face of it we would expect the earliest wheels to be of a single piece of wood similar to those so often depicted by humorists, in which case the wheel may well have been developed initially outside Mesopotamia, even perhaps as far away as on the steppe-lands of Asia. That this may have been so is to some extent suggested by the fact that solid wheels made of a single piece of wood do turn up in Western Europe at a very much later date. Indeed it seems likely that in Europe the wheel was acquired by diffusion from some centre other than Mesopotamia as, for example, Southern Russia.

With the introduction of the shoulder-yoke rather than attachment to the horns it became possible to use the onager as a draught animal, but the bridle and bit were still unknown as a means of guiding it and the reins were attached to a ring of copper passed through the animal's nose – similar to the ring still used today for controlling bulls. The yoke was conceived as a cross-member to a single draught pole and this meant that for any vehicle there had to be a pair of animals. Furthermore, the onager's shoulders, unlike those of the ox, are not ideally suited for a yoke, and it became necessary to pass a collar around the animal's throat to keep the yoke in place, as well as a girth-band to keep the yoke down on the animal's shoulders. The effect of the breast-collar was to partially choke the animal, particularly if it exerted itself greatly, and this tended to reduce its efficiency. The onager, anyway, was a fairly small animal and therefore it is not uncommon to find not two but four onagers used, the outside pair being attached by their collars only to the two ends of the yoke. The introduction of paired draught was in the long run to have a very considerable effect upon the history of the development of wheeled transport, for it seems that once the use of the single draught pole and the yoke was established, the intellectual effort required to change this means of traction was too great, and indeed within the span of this particular history we shall barely see the introduction of either the shaft or the horse-collar. Thus in the ancient world there was no such thing as efficient, heavy horse-transport.

It is quite probable that the introduction of paired draught was the outcome of using beasts of burden to pull ploughs, for there is some evidence of ploughing long before any form of wheeled vehicle, though since ploughs were constructed entirely of wood we have no early examples remaining from excavated sites. It is nevertheless interesting to notice that while in many of the older civilizations the invention of the plough was attributed to a god the introduction of the wheeled vehicle does not appear to have received the same divine attribution. The earliest ploughs, anyway, were probably pulled by a team of men and women, and indeed in some of the earlier Egyptian tomb paintings we have pictures of just such an operation.

The earliest form of plough that we know of amounts to little more than a forked stick, the two tines of the fork serving as handles and the junction as the share. Initially a rope was attached just above the junction and on this rope the plough team pulled while the ploughman pushed down on the handles. The effect was to produce a rather narrow furrow in the soil. Before ploughing could begin the soil had to be broken up, and although most of the soils cultivated at this time were fairly light, we find not only that the hoe was used for this purpose but also that the larger clods of earth were broken up by mallets. Into the furrow made by the primitive plough described above the sower put his drill of seed. In time the plough's tow-rope was replaced by a draught pole secured to a yoke at one end and lashed at the correct

77 Pictograms of ploughs from Egypt of about 3000 B.C. with a reconstruction based on them and on later pictures of ploughs

In both Egypt and Mesopotamia early forms of plough were little more than a forked branch dragged through the soil by a pair of oxen, the ploughman holding the two branches of the fork as handles, while the junction, sharpened to a point, served as a primitive share. The ploughs have long since disappeared, but their shape can be deduced from the pictograms used to record them in both areas, as well as from later pictures (Figure 119).

angle to the V-fork at the other. Indeed, this type of plough survived with remarkably few modifications almost throughout the history of ancient Egypt, although the lashings between the draught pole and the share were ultimately replaced by wooden cross-braces while the handles were often braced with wooden struts. Although early Mesopotamian ploughs appear to have been of this kind, by about 3000 B.C. we find the introduction of the first major modification. A single pointed piece of timber formed a share and sole – the share actually cutting the soil while the sole helped to push it aside, so creating a deeper and wider furrow.

Although by our standards these early ploughs must be seen as hideously inefficient, for in practice they failed to do more than just scratch the surface of the soil, nevertheless they must have resulted in vast improvements in crop yield, for not only did they allow sowing to become consistently even, but since the crops were in regular rows they allowed efficient weeding. Of all the devices created by mankind up to the end of the fourth millennium B.C. there can be little doubt that the plough had overall the greatest effect. More probably than not it was mainly responsible for the rise in population in the small Mesopotamian and Egyptian cities.

4
The Early Dynasties

(3000–2000 BC)

The two or three centuries centred around the year 3000 B.C. appear to have been fairly critical in the history of the development of early technology. This view may be fortuitous, for not only from this period do we begin to get Mesopotamian written records that we can comprehend, but also the political and religious situation in both Mesopotamia and Egypt gave rise to elaborate forms of burial, and from the many tombs known to us we can derive a vast amount of information. The rulers in these countries as well as the more important citizens elected to be buried in capacious graves with all the goods and chattels that they imagined they might require in the next world, while at a slightly later date there began the series of complex tomb paintings in Egypt that were to furnish us with so much valuable information. So what seems to us to be an era of sudden technological advance may appear so only because we have a dramatic increase in sources of information.

In the lower Mesopotamian Valley there had evolved two small kingdoms. In the immediate area of the delta were the Sumerians, of whose cities Ur is probably the best known, while further to the north were the Akkadians, occupying an area in the Euphrates Valley with Babylon as its approximate centre. In both Sumer and Akkadia there were several cities, each with its own ruler, and while the two kingdoms were often at war with each other in neither was political unity achieved. Nevertheless both regions evidently became extremely wealthy, and this wealth must have depended entirely upon the products of agriculture and husbandry, for apart from their crops and livestock and an almost inexhaustible supply of mud, the countries were quite devoid of natural resources, all other materials of necessity having to be imported. Even so, from the graves of the rulers of these small cities have come many objects which show that considerable technical advances were being made.

In the Valley of the Nile a somewhat similar development appears to have taken place, but even before 3000 B.C. we find that the cities had formed a loose political alliance and that the country as far upriver as Aswan was divided into two major units – the one, Lower Egypt, that area of the delta and the Nile Valley as far south as Memphis; and the other, Upper Egypt, between Memphis and Aswan. The two countries were ultimately to be unified by the ruler Menes, the first Pharaoh, who established his capital at Memphis. Tradition has it that Menes not only concerned himself with the unification of Egypt but also with the control of the river, and to him is attributed the first damming of the Nile, the digging of dykes for agricultural purposes, and indeed the first attempt to control and apportion the waters of the river. The wealth of Egypt was thus, with Mesopotamia, based upon its agricultural output. However, unlike Mesopotamia, the Egyptians had on their doorstep a number of mineral resources that they were able to exploit with little effort, including copper ores, gold and a wide range of rocks suitable for building and the making of a great variety of ornaments.

Somewhere shortly before the year 3000 metallurgists made a discovery that was to transform the entire industry. They found that by mixing a small quantity of tin ore with the copper ores when they smelted them a harder and altogether more useful metal than copper was obtained. Briefly, they discovered the alloy, bronze. The occurrence of tinstone can never have been very widespread in the Near East and it is interesting to notice that geologically speaking it does not occur in the same type of deposit as do the ores of copper, but rather it occurs in those areas in which we might expect to find veins of gold. Thus, tinstone which is a fairly dense mineral may well have first been noticed during washing for gold, and metallurgists finding that the little black lumps of ore were relatively heavy presumably made various attempts at smelting them until they arrived empirically at a suitable alloy with which to make tools and weapons. Bronze, however, was not only a harder material than copper but was also more easily worked, for the effect of alloying a small quantity of tin with copper is to reduce the melting-point of the resulting metal, and in practice this meant that at the same temperature at which they were accustomed to cast copper the bronzesmiths found now that they had a far more fluid metal which was much easier to cast. With the introduction of bronze, therefore, the quality of casting improved dramatically.

It is most unlikely that this discovery was first made in Mesopotamia, and in all probability it was made much nearer the sources of the metals, in the mountainous areas of Syria or Eastern Turkey for example, but the Mesopotamians had the wealth to purchase this new metal and the wealth to employ craftsmen to fashion it. Thus it is from the tombs of the early Sumerian kings that we find the first examples of bronze used in any quantity. The situation in Egypt was totally different. While endowed with large quantities of copper ore the country was utterly devoid of tinstone, and thus at this early period the Egyptians continued to use copper where in Mesopotamia bronze would have been employed, and it was not for another thousand years that bronze became the common metal in Egypt. But despite the difficulty of working the pure metal the Egyptian smiths achieved remarkable results with copper alone.

A study of the metal objects from the royal Sumerian tombs shows that the smiths had made very considerable technical advances. Complex objects were often cast in moulds made of two, three or even four pieces and it is quite clear from the chemical composition of some of the weapons and ornaments that the smiths had been experimenting quite widely with various alloys to such an extent that they had hit upon the idea of joining one piece of metal to another with an alloy of a different composition and had thus in effect begun to experiment with solders. But excellent though their technique of casting was, much of the work was still accomplished by laborious hammering and chasing, while the size of the castings they were able to make was still severely restricted and seems to have been limited by the quantity of metal that they could melt at any one time. Nevertheless, some fairly large pieces of ornamental metal-work were produced by the extremely simple expedient of beating sheet copper over a wooden substructure to which the sheeting was attached by nails.

We saw in the last chapter how early carts and wagons were unlikely to have played a large part in the long-distance transport of imports into Mesopotamia, and it is now for the first time that we get any direct indication of the type of boat being used for this purpose in Sumer and in Egypt. In both areas the boats appear to have been made largely of bundles of reeds laid horizontally and lashed together. From Mesopotamian seals and their impressions, as well as a model boat in silver from Ur, it is evident that here the early vessels were built of bundles of reeds terminating with upturned, pointed bow and stern. Indeed the silver model just referred to could equally well be one of a boat used today by the Marsh Arabs in the delta of the Euphrates, which are built in exactly this manner. From Egypt we have one or two very bad pictures of early boats drawn on pottery from the pre-dynastic period,

78

79

80

78 Boat depicted on a jar from Egypt, before 3000 B.C.

79 Men building a reed boat, from an Egyptian tomb painting about 2500 B.C.

80 Boat depicted on a stone vase from Mesopotamia about 3000 B.C.

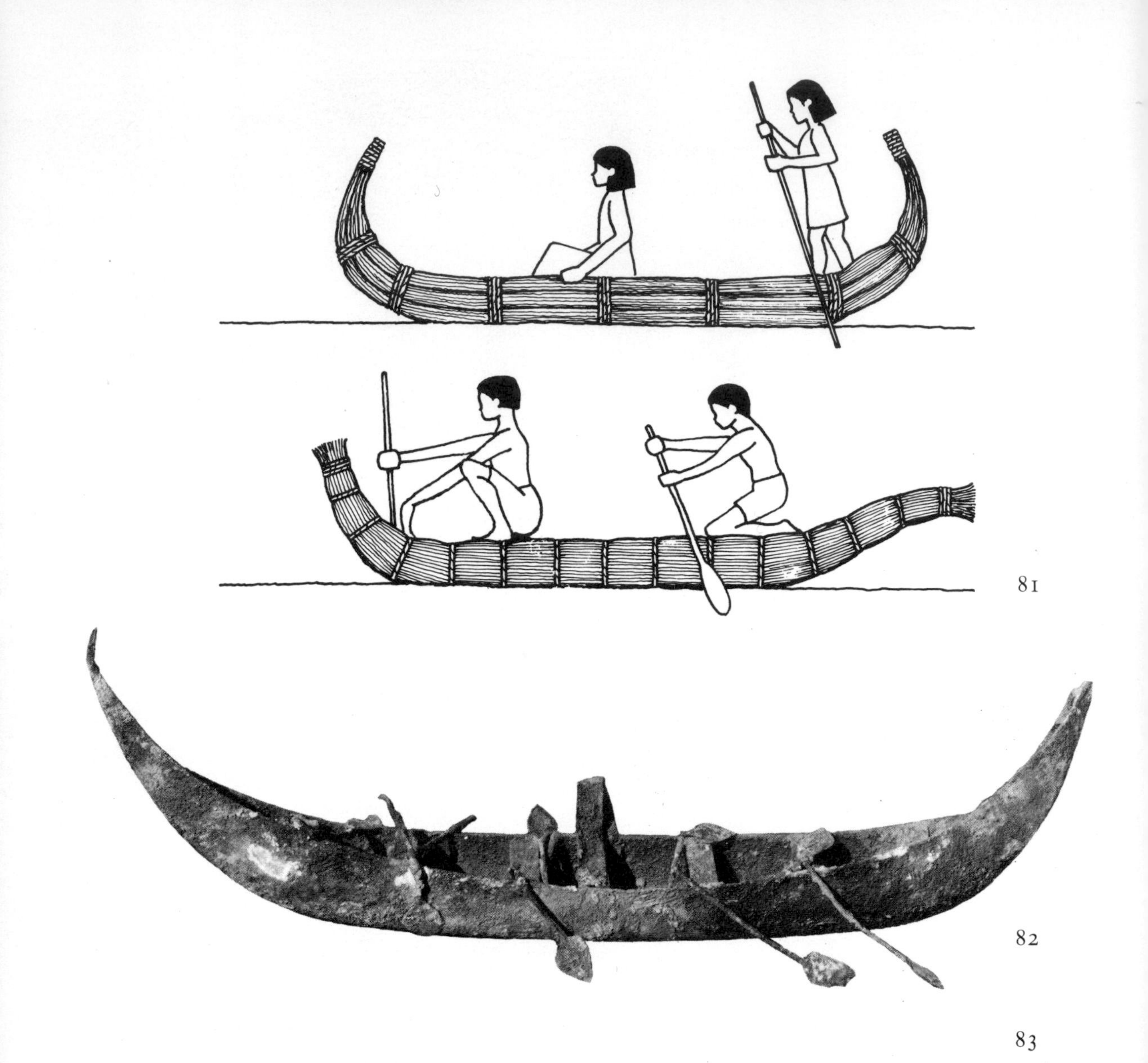

81

82

83

81 Reconstruction comparing Egyptian and Mesopotamian boats of about 3000 B.C.

82 Model of a boat in silver, Mesopotamia about 2500 B.C.

83 Boat used by the Arabs in the Euphrates delta today

Well before 3000 B.C. pictures of boats appear on Egyptian pottery and on Mesopotamian seals, but without a knowledge of what was to happen in the following few centuries it would be impossible to guess what they were like or of what they were made. The Egyptian jar shows a boat with an exaggerated number of paddles, and perhaps a square sail.

The much later silver model of a boat from Ur, however, gives a far clearer picture of this type of craft, while the Egyptian relief of much the same date leaves no doubt as to how and of what such craft were constructed. Reeds were tied into bundles, and then were further lashed, bundle upon bundle, to produce a boat with no keel and upswept stern and bow.

The major structural difference between boats in Egypt and Mesopotamia appears to have been in the treatment of the bow and stern. In the former region the terminals were left blunted, the cut end of a single bundle of reeds; in the latter area the bow and stern were brought to a point. In all probability the Mesopotamian craft were covered with bitumen as are the boats of the Marsh Arabs today.

(before 3000 B.C.), but these drawings become intelligible only when we compare them with the really superb tomb paintings of the later periods. The Egyptian boats were clearly built of bundles of papyrus, and although both bow and stern were raised the bundles of reeds were cut off to give a blunted, upturned end. In both regions the boats were of course without keel and of shallow draught. They were also light, and this may have been an important factor in negotiating the rivers, for at the cataracts portage would have been essential.

In both Mesopotamia and Egypt the boats appear to have been paddled at the beginning of the third millennium B.C., and it was only later that the Egyptians took to rowing their vessels. The very poor drawings from the pre-dynastic period in Egypt already mentioned would suggest that some of the early boats were being sailed, but at this stage of development a single mast evidently proved too difficult to step, and we find in use a bipod mast supported by an after-stay rope. The sails were presumably of linen, each vessel having a single square-rig sail which was furled by lowering the yardarm.

Vessels constructed along these lines obviously had very severe limitations: if they were built too long they would clearly break their backs in rough water or when heavily laden: they had little freeboard and would thus easily capsize in a rough sea. They were, hence, essentially river-going craft and it is not for another five hundred years that we shall find anything approaching a sea-going ship. Nevertheless, for the carrying of valuable materials in both Mesopotamia and Egypt there can be little doubt that the boat was the most important single means of transport. Heavier loads of course could have been shifted by raft, and although we may surmise that rafts were already in use – indeed reed boats were most probably developed from rafts – it is not until we come to the period of stone monumental buildings that we find reliefs showing the carrying of huge blocks of stone by raft.

The very different nature of the available raw materials in Mesopotamia and Egypt led to strongly contrasting methods of building in

84 An Egyptian vase showing a boat with sail, about 3000 B.C.

By 3000 B.C. reed boats of the kind just described were being sailed by the Egyptians. A simple square sail, probably of linen, was set well forward. Very early pictures show a single mast, although later this was replaced by a bipod mast.

85 Building a wooden boat, as depicted on the walls of an Egyptian tomb of about 2500 B.C.

By 2500 B.C. the Egyptians were certainly building boats of wood, possibly for many centuries before. But their method of construction was clearly inspired by the reed boat: planks were dowelled or sewn, edge to edge, without keel or ribs.

the two areas. In Mesopotamia, as we have already seen, brick, at least for the more important buildings, was already being fired, and we now find in common usage a type of brick apparently evolved to overcome a particular difficulty. A simple flat mud-brick tends to warp on drying and to bend still further on firing. The Mesopotamian brick-makers developed the trick of building up a dome of mud in the top of the mould which prevented excessive warping but which gave a brick which was flat on the underface and convex on the upper surface. Using these bricks the builders experimented with a wide variety of bondings including a herring-bone arrangement, and it was presumably experimentation of this sort that led to the development of the true arch which first appears at this period, thus enabling builders to span wide spaces without the need to import large stone lintels. The Sumerians, of course, did import a limited amount of fine stone for facing and for querns, but their buildings remained, nevertheless, essentially brick ones.

By contrast, the Egyptians were content to continue building with unfired mud-brick for all domestic purposes including royal palaces, and because of this we know less about their normal domestic lives than we would do otherwise, for these buildings have long since been totally destroyed. For monumental building, however, they began to quarry and to dress sizeable blocks of stone. Thus, the pre-dynastic kings and those of the very early dynasties were buried in mud-brick houses of the dead surrounded by all the objects that they felt they might need in their afterlives. These mud-brick structures were too easily robbed, and to overcome this desecration the burial chamber was then sunk beneath the house which itself became a completely stylized copy made of mud-brick and rubble, although the burial chamber and the exterior of the structure were sometimes faced with stone. Ultimately this arrangement of sunken burial-chamber surmounted by a stylized house of the dead was to lead to the design of the pyramid as a suitable mausoleum for the Pharaoh and his family.

The stone most favoured for building pyramids was a fine limestone obtained from one of a limited number of quarries in the vicinity of the Nile itself. By this time the Egyptians had become fairly adept at working stone. Even in pre-dynastic times they had learnt how to make stone vases and other vessels out of harder rocks than the limestone of which the pyramids were constructed, so that as a building material it presented few technological problems. Each block of stone appears to have been cut clear of the parent rock at either side and at the back by making a narrow draught, using either copper chisels and mallet or hand-held stone picks made of diorite. Along the line of what was to be the lower edge of the block a series of pits were then

86 Quarrying wedge slots, Aswan

Stone for building pyramids was quarried by the same method as that still in use today, although the tools were all of stone, wood and copper. A line of slots was cut into the quarry face with stone picks or copper chisels. Wooden wedges were then driven into the slots, causing the rock to split along the line.

cut and into these were driven wooden wedges, causing the block to split away from the parent rock. It should perhaps be pointed out that even today, although mechanical tools are used for much of the cutting, the same method of quarrying is still used in many parts of the world.

For the Egyptians a far greater problem was presumably presented in the shifting of these blocks of stone once quarried, for many of them weighed several tons. To transport the building stone down the Nile by barge was a relatively simple matter; the principal difficulty must have been to carry the stone from the quarry to the river and from the river to the building site. Wheeled vehicles never became popular in Egypt until very much later, although the Egyptians of this period occasionally used wheeled assault-ladders as a military device, as we shall see later. For transporting heavy loads overland the Egyptians seem to have preferred to use a sledge, and there are several tomb paintings showing the way in which these heavy burdens were moved. Occasionally the sledges were pulled over rollers, but this was not invariably the case, and more often one feels that the future trackway for the sledge was carefully prepared beforehand. Water poured in front of the runners was used as a lubricant and a large team of men hauled the sledge by massive tow-ropes usually made of papyrus.

Once the sledge had arrived at the building site there remained the problem of lifting the block of stone into position, but on this subject the tomb paintings give us no help. However, from various excavations we have sufficient evidence to suggest that a ramp was built against the side of the pyramid up which the sledge and its burden were drawn. The block of stone was ultimately levered into position from the sledge and as the pyramid grew, so did the ramp which was removed after the last block had been laid. Although before being lifted each block of stone was trimmed carefully to fit precisely to its neighbours, the ultimate face-trimming was not carried out until after the building had been completed. For this work it seems highly unlikely that the mallet and chisel were used, but rather that it was done with hand-held picks of diorite or other tough stone, while for a really fine finish the surface would have been rubbed down with blocks of sandstone.

Although from the year 3000 B.C. onwards bronze and, in Egypt, copper, were used increasingly for the making of tools, one has to confess, nevertheless, that a large part of the production of these metals went into the manufacture of arms and weapons. The reliefs and wall paintings of this period obviously show an idealized situation, but even allowing for that, it is perfectly clear that the rulers of the various

87

88

87 A relief showing a phalanx of troops from Mesopotamia, about 2500 B.C.

88 A tomb-model showing a phalanx of troops from Egypt, about 2000 B.C.

The power of both the Egyptian and Mesopotamian rulers depended to a large degree on well equipped and disciplined armies. For the rank and file, shields and helmets were made of leather. The main weapons were the spear and the arrow, both of which were most commonly tipped with stone points. Since only the most wealthy could afford metal arms, copper and bronze were reserved largely for officers and crack troops. The armies were composed almost entirely of infantry – spearmen and archers.

states had well-equipped and well-trained armies. The backbone of the army was the infantryman who was usually armed with a spear, a dagger and a large shield. Also many of the infantrymen wore helmets, but it is unlikely that these, or the shields of the rank and file, were made of metal; more probably they were made of leather. The Sumerians and Akkadians, but not the Egyptians, also employed a number of wheeled vehicles, but it is far from clear what their function was in battle. Wagons with four solid wheels drawn by four yoked onagers seem to be mobile armouries rather than chariots, and were probably not used in the fore of the battle at all. But there are other lighter two-wheeled vehicles, equally drawn by four yoked onagers, that may possibly have been used as chariots. Even so, their design suggests that they were fairly slow, lacked manoeuvrability and were exceedingly vulnerable, and one can well imagine that well-trained infantry would soon have learned to cope with them.

Apart from ensuring that one had an army as well trained and equipped as one's enemy the principal method of defence was to gird one's city with an adequate wall. In fact, city walls appear far earlier than this period. Even the pre-agricultural community at Jericho had set a defensive wall around their village, and many of the early agricultural villages were designed in such a way that their defence from sudden attack could have been an easy matter. Now, however, the tall city wall became *de rigueur*, while its limited number of gateways allowed not only a manageable defence but also a means of regulating the comings and goings into the city. With one's enemy safely protected by the city wall the attacking army had little choice but to attempt an assault or to settle down and besiege the city. At this early period the only assault vehicle appears to have been a scaling ladder on wheels – known from an Egyptian tomb painting – and methods of breaching

89 Copper model of a chariot from Mesopotamia, about 2500 B.C.

90 Reconstruction of this type of chariot based on the copper and contemporary clay models

The Mesopotamians were experimenting with chariots, however. These were drawn by four onagers, the inner pair yoked, the outer pair pulling on traces. The vehicles seem unduly cumbersome and, one feels, the infantrymen could have easily dealt with them. Probably they were used in battle to give the commanders greater mobility.

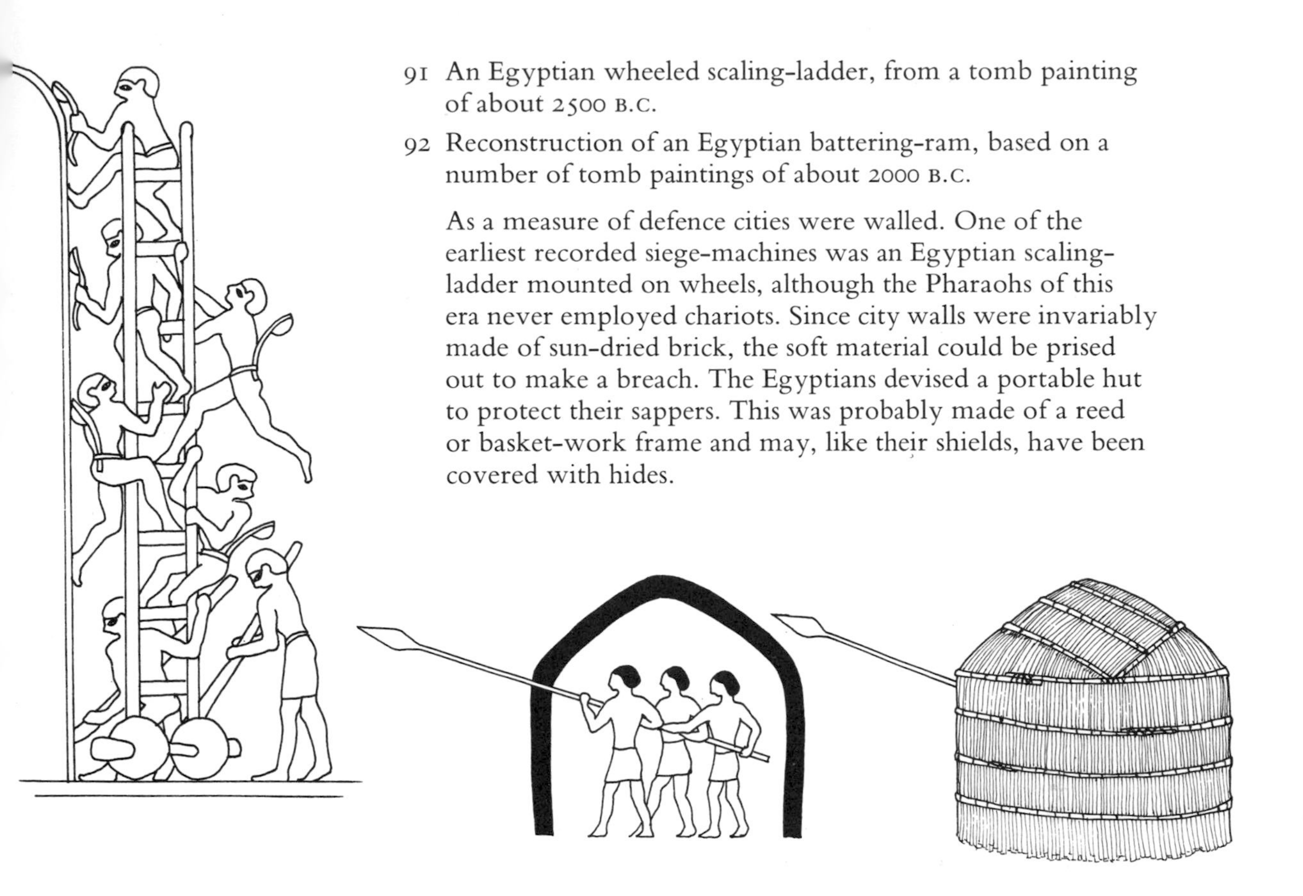

91 An Egyptian wheeled scaling-ladder, from a tomb painting of about 2500 B.C.

92 Reconstruction of an Egyptian battering-ram, based on a number of tomb paintings of about 2000 B.C.

As a measure of defence cities were walled. One of the earliest recorded siege-machines was an Egyptian scaling-ladder mounted on wheels, although the Pharaohs of this era never employed chariots. Since city walls were invariably made of sun-dried brick, the soft material could be prised out to make a breach. The Egyptians devised a portable hut to protect their sappers. This was probably made of a reed or basket-work frame and may, like their shields, have been covered with hides.

the wall do not seem to have been effective. Unless the city garrison were very small, one can only surmise therefore that the more normal course under these circumstances was to besiege the city.

In the two or three centuries immediately following the year 3000 B.C. the technological innovations that had taken place in Mesopotamia and Egypt were being slowly diffused into other areas of the Near East. Copper was being widely used in Anatolia and also in Cyprus and in Crete, and while bronze appeared just as early in Anatolia as it did in Mesopotamia it was not until shortly before the year 2000 B.C. that it became commonly used in this area. The introduction of the knowledge of copper-working into Cyprus, however, was to have very considerable repercussions, for the copper ore deposits in that island were by any standards fabulous and a busy trade in the metal between the island and Anatolia and Syria was rapidly built up. At this point in time, too, the city of Knossos in Crete appears to have been developed as a trading centre, reflecting architecturally many of the aspects of the cities of the Near East. As we have said, its contacts must have been entirely by sea, and it is more than just mildly infuriating that early pictures of ships recovered from this site are so inadequate that they allow us to make no reasonable reconstruction. There must

93 A sketch of a ship from a fragment of a vase, and a clay model from Crete, about 2000 B.C.

The development of shipping in the Eastern Mediterranean was quite different from that in Egypt and Mesopotamia. Early pictures of ships from Crete are as frustrating as those from Egypt, but they do show that the vessels were low in the water with a high, upswinging stern.

94 A conjectural reconstruction of a Cretan ship of about 2000 B.C.

This reconstruction of an early Cretan vessel is based partly upon pictures of the period, and partly upon fishing vessels still being used in Senegal today. The major part of the hull is a hollowed log brought to a projecting point in the bow to help when beaching. The sides of the vessel are built up from a single line of planking, while a curved post and further planks form the stern.

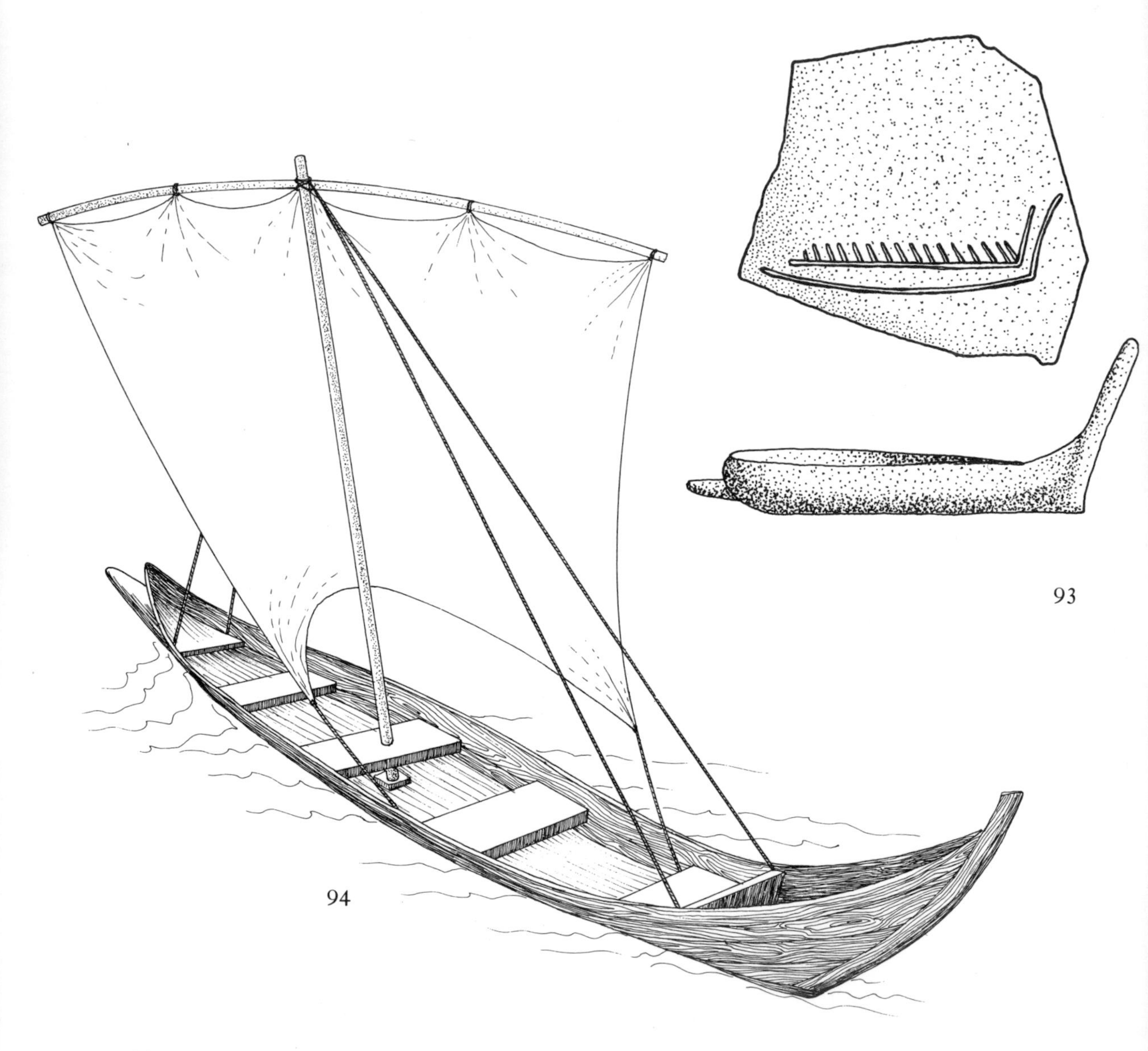

95 Building a fishing-boat, East Africa

The type of construction suggested for early Cretan shipping can be seen clearly in this photograph. In the foreground the inverted hull is being shaped by adze (compare Figure 85), while the projecting prow, here a distinct timber, and the additional planking forming the sides of the vessels can be seen in the boat drawn up on the beach.

96 Pictogram of a plough from Crete, about 2000 B.C. and a reconstruction based on this and on ploughs still in use in Western Anatolia

Many aspects of life in Crete can be attributed to her contact with countries overseas. For example, the Cretans used seals similar to those in Mesopotamia and kept records on clay tablets. They developed their own script, however. One early pictogram shows a plough, but it differs completely from those in use in Egypt and Mesopotamia, for it had only a single handle, leaving the ploughman a free hand with which to goad his oxen. This type of plough may well have been an import from Greece or Anatolia, but we have no record of equally early ploughs from these areas.

97

by this period have been some viable timber-built shipping capable at least of coastwise traffic, working the northern shores of the Eastern Mediterranean. But the form and dimensions of these ships escapes us completely, and we shall find that the nature of shipping in this area will be a recurrent problem for the historian throughout the period covered by this book.

So far, we have not been able to say a great deal about many crafts, especially those dealing with perishable materials, but now with copper and bronze tools, with wall paintings and engravings, and with better preserved finds from archaeological excavation we begin to see just how far many crafts had developed. For example both in Mesopotamia and in Egypt we now find properly jointed furniture, although this could probably only have been afforded by the very wealthy citizens. The axe and chisel both previously existed as flint or stone tools, and we now find them made of copper and bronze, although it is interesting

97 The sledge of Queen Shub-ad, from a Mesopotamian tomb of about 2000 B.C.

98 Egyptian carpenters of about 2500 B.C. as depicted on the walls of a tomb

Woodworking now became very sophisticated. Axe and adze were used for rough shaping; timber was lashed between upright supports while being sawed; mortices were cut with mallet and chisel; a final polish was given with blocks of sandstone. Because so little wood has survived from Mesopotamia it might appear that the craft was not so well developed as in Egypt. In fact, rare examples of woodwork, such as this sledge from the tomb of Queen Shub-ad, show that the Mesopotamians were every bit as advanced.

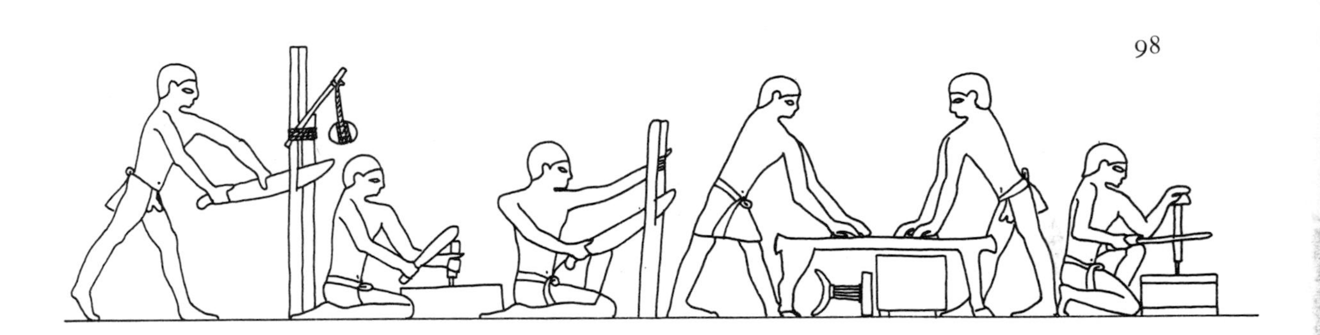
98

to note that as a carpenter's tool the adze seems to have been more popular than the axe. Wood was cut with a saw which resembled if anything a rather large kitchen knife with a toothed edge, and from surviving examples it is clear that there was no attempt to set the teeth as with a steel saw. Often the wood being cut was held in a simple vice made by driving two posts into the ground and lashing the timber to be cut between them or by lashing the timber into the split of a cleft upright post. While smaller holes were drilled with a bow-drill, the mortises and carving were done with a wooden-handled chisel struck with a wooden mallet, almost identical in shape to the wooden mason's mallet still used today.

An enormous industry in stone vases developed in Egypt, and while some of the vessels were made of the softer materials such as calcite the Egyptians were perfectly able to work granite and other hard rocks. The vases were roughed out in the solid from a block to give a cylinder approximating to the shape of the vase. For hollowing out the cylinder a special type of drill had been developed. At the lower end of an upright shaft was fixed a crescentic piece of flint which formed the bit, while the upper end of the shaft was fitted with a cross-handle which was weighed down by large boulders contained in string bags attached to

99

99 Stone bowl from an Egyptian tomb of about 2500 B.C.

100 Hollowing stone vases: a scene from the walls of an Egyptian tomb of about 2500 B.C.

101 A reconstruction of the drill used to hollow stone vases

In Egypt, where there was an ample supply of fine decorative stone, there developed a large industry devoted to producing really superlative stone wares. The vessels were first roughed-out by flaking off the unwanted material from a suitable block, but the final shaping and polishing was done entirely by rubbing with sandstone. The only specialized tool introduced into the industry was the drill for hollowing the stone-blocks. The drills were fitted with a crescentic bit of flint turned by hand, and weighed down by rocks attached to the handles.

100

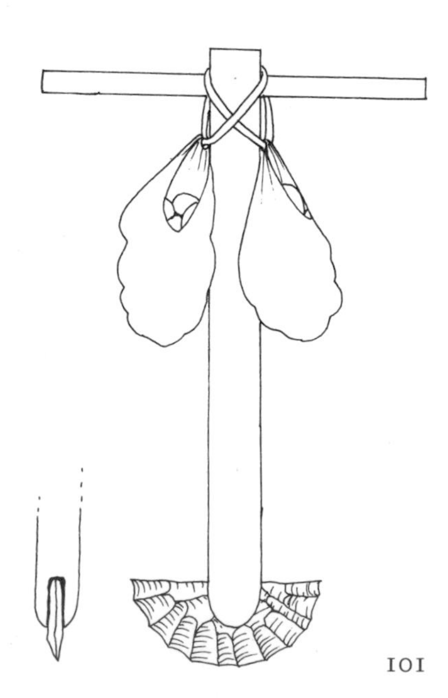
101

it, the whole device being rotated by one or two workers. The final surface finish of the vessel was made by polishing with a sandstone rubber, and indeed the same type of sandstone rubber was used by carpenters for finishing off their woodwork.

Like their Mesopotamian counterparts the Egyptian rulers found it necessary to keep records, but instead of using clay tablets as was the case in Mesopotamia the Egyptians devised a lighter and more convenient way of storing information in the form of papyrus rolls. The papyrus reed, of course, grows prolifically in the Nile Delta, and we have already seen how it was used for making boats and in the manufacture of ropes. Strips of this reed were laid edge to edge on a flat surface and further strips were laid at right angles to them and across them, again, edge to edge. The material was then beaten hard with a heavy wooden mallet, causing the strips to fuse and to become flattened, so forming a paper-like material that could be rolled up for storage. This was, of course, an easy material on which to write with ink, and although the Egyptians started their system of recording with pictograms, as did the Mesopotamians, there was no real necessity to drastically modify this method of writing. Thus their pictograms slowly changed their meaning and became totally stylized to give the familiar hieroglyphs, but throughout the history of ancient Egypt there was to be no other great change in methods of writing and recording, save that very early the Egyptians reduced their hieroglyphs to a highly simplified cursive form that could be written rapidly by hand.

102 Two men beating out papyrus: from an Egyptian tomb of about 2500 B.C.

Unlike the Mesopotamians, the Egyptians seldom kept their records on clay tablets. Instead they wrote in ink on papyrus. To make this material strips of the papyrus reed were laid on a flat surface and further strips were laid across them at right angles, and the whole was then hammered until all the strips welded to form a continuous sheet. For storage the papyrus was rolled in convenient lengths.

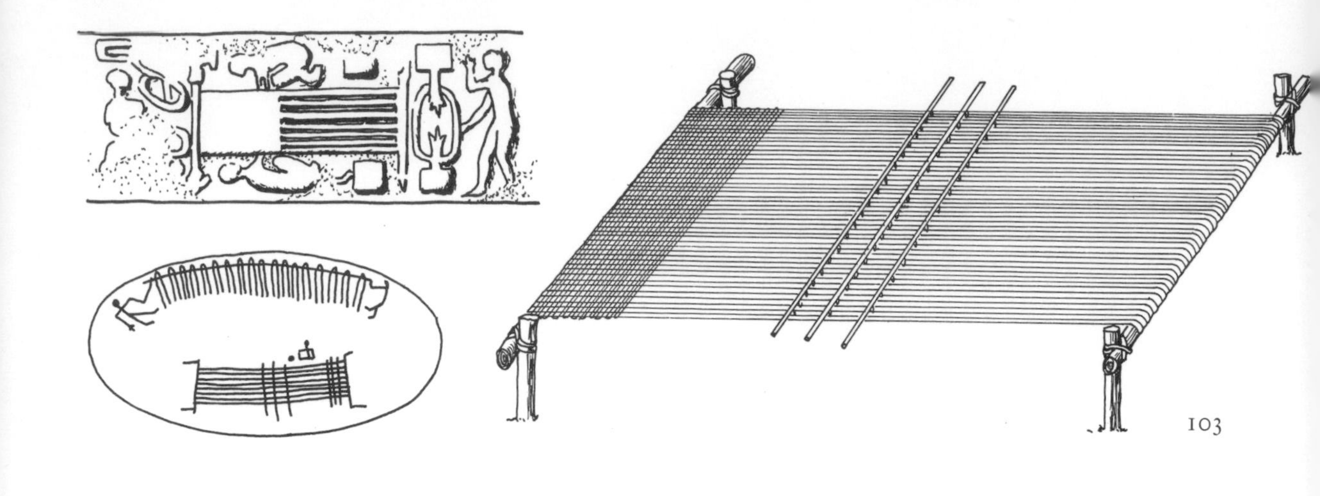

103

104

105

103 Two early pictures of looms from a Mesopotamian seal and an Egyptian painted dish, both before 3000 B.C., with a reconstruction of this type of loom

104 Horizontal looms from an Egyptian tomb painting of about 2000 B.C.

105 A horizontal loom used by Andean peasants today

The earliest pictures of looms appear shortly after 3500 B.C. on a seal from Mesopotamia and a painted dish from Egypt. Both were essentially of the same type: the warp threads were stretched between two beams which were lashed to short upright posts at no great distance from the ground. The Mesopotamian seal gives no impression as to how the weaving was done, but the Egyptian dish shows three rods lying across the warp. This may mean that three different lines of weave were being produced. In other words a step may have been taken towards making pattern weaves, rather than plain weave, on the loom.

Pictures of looms of the period 3000–2000 B.C. from Egypt show the same type of loom in use, although as an artistic convention they are invariably shown in false perspective as though rotated through a right angle. Looms of this type are still in use today in many remote parts of the world.

Cloth continued to be made on a very simple form of loom in which the warp threads were stretched between two posts, held a short distance horizontally from the ground by forked pegs. Into this warp was woven the weft that had been wrapped round a bobbin. For their finer raiment the Egyptians used linen – the flax plant being found native in the Nile Valley. Wool on the other hand was not much favoured by the Egyptians although they did make cloaks of the material. But further north, particularly in Anatolia, such little evidence as we have would suggest that nearly all cloth of this period was made of wool.

The date-palm and barley provide us with the first direct evidence that we have that both the Mesopotamians and the Egyptians were making fermented drinks. Date-palms grow prolifically in both areas, the fruit themselves are rich in sugar and in these warm climates fermentation is fairly rapid. Thus while the first proof of the making of date wine comes from the period there is little doubt that it was being prepared by the people in these areas long before this time. Since the fruit itself normally contains the micro-organisms necessary to cause fermentation the making of date wine was not of itself a very complex process, and fundamentally one only had to have a large jar in which to put the date mash while fermenting and some device with

106

which to strain it once fermentation had taken place. The making of beer on the other hand was a far more sophisticated process, for some of the cereal, usually barley, had first to be allowed to germinate, and in so doing much of the starch in the grain was converted into sugar so making the ultimate drink both palatable and slightly sweet. Hence, the first stage was to moisten the cereal and allow it to germinate for a short time. The malted cereal was then made into loaves and baked lightly. These malted loaves were then crumbled and added to water with more grain and this mash was allowed to ferment for three or four days. It was then strained, bottled and stoppered to prevent further fermentation which would have made the drink acidic.

At this period, too, the grape was also being cultivated and the fruit converted into wine, and apart from scenes depicting the making of wine we also have records of vineyards. As far as one can gather, these vineyards were the property of the rulers both in Mesopotamia and in Egypt, and it seems unlikely that the common man could ever hope to drink wine made from the grape. It has even been suggested that grape wine was reserved entirely for the gods. But the actual process followed seems to have been more or less that as we know it today. The grapes were harvested, trodden, allowed to ferment; the juice was then strained and bottled. Sometimes, too, grape or date wines, and even the beer, were flavoured with herbs; in many records we read of more than one variety of date wine and more than one variety of beer, and it would seem that the additions made to these drinks and the precise methods of manufacture were kept as jealously guarded secrets – perhaps the first manufacturers' secrets that we have come across in this history of technology.

Apart from the grape other fruits and some vegetables were being brought into cultivation, and the monarchs both in Mesopotamia and Egypt seem to have prided themselves on the variety of plants growing in their gardens, while considerable effort appears to have gone into

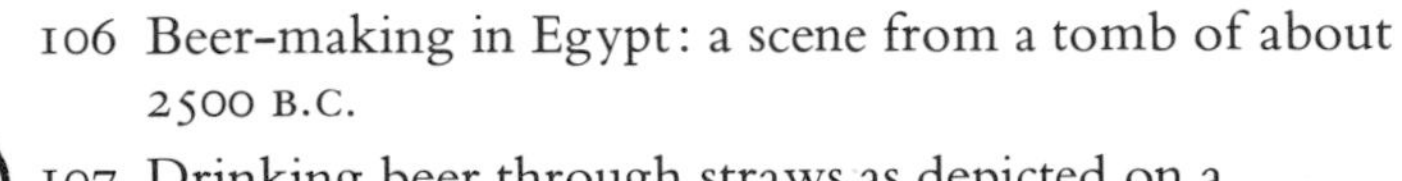

106 Beer-making in Egypt: a scene from a tomb of about 2500 B.C.

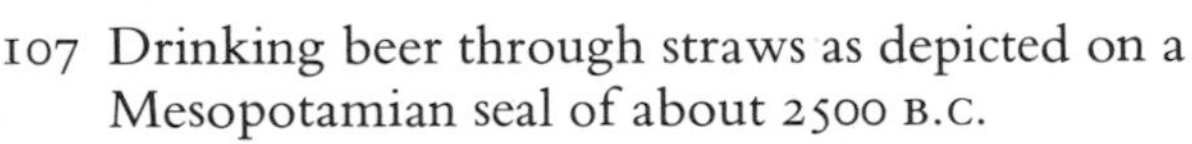

107 Drinking beer through straws as depicted on a Mesopotamian seal of about 2500 B.C.

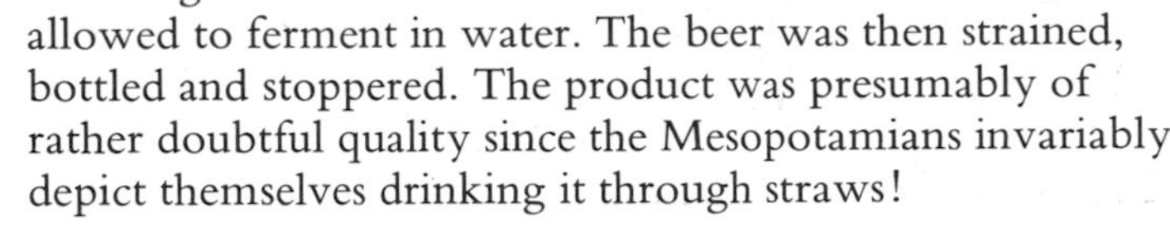

Beer was being brewed in both Egypt and Mesopotamia. Malted grain was first made into loaves which were then allowed to ferment in water. The beer was then strained, bottled and stoppered. The product was presumably of rather doubtful quality since the Mesopotamians invariably depict themselves drinking it through straws!

107

108 Early Egyptian grape presses from a tomb painting of about 2000 B.C.

109 A reconstruction of the Egyptian bag-press

Both the date-palm and the grape were used for making wine. To extract the juice from grapes a simple press had been devised. The grapes were put in a tall linen bag which was then stretched between two posts. Bar handles were placed at both ends of the bag and twisted.

108

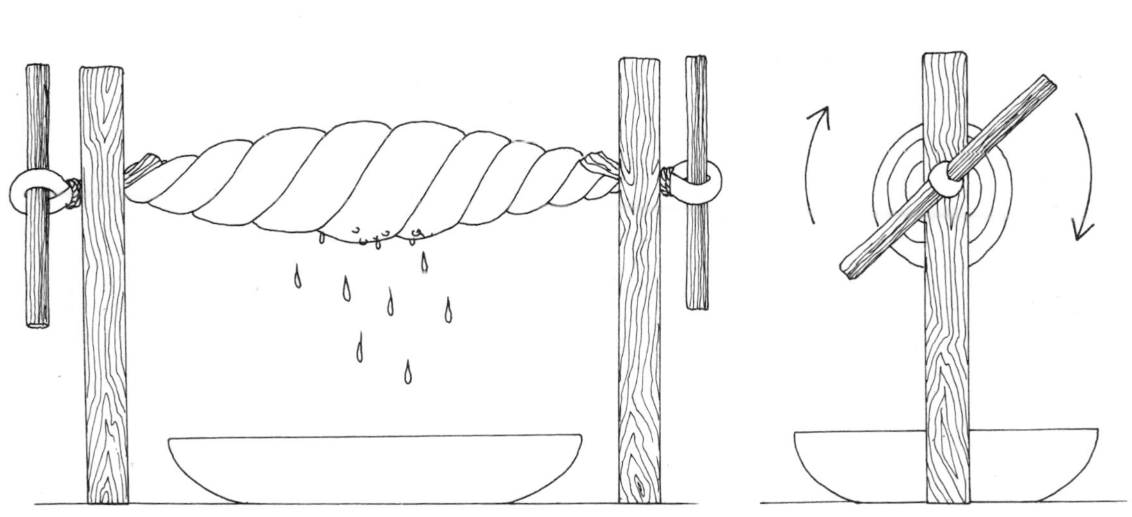

109

110

111

the establishment of exotic fruits and vegetables. Naturally these gardens required watering all the year round, and although we may be sure that in Mesopotamia perennial irrigation had already advanced considerably before this period, in both Mesopotamia and Egypt we have direct evidence of a water-lifting device, the shaduf. Essentially this was a beam pivoted into the top of an upright post. At one end of the beam was attached a rope on the end of which was the vessel to contain the water, while at the other end of the beam was a counter-weight. The vessel was lowered by hand into the water of the river, raised, emptied into either a cistern or into an irrigation channel and then lowered again into the water. The shaduf, of course, took a great deal of the backache out of raising water which before this time, presumably, had to be done entirely by hand, but it remained a slow and tedious job. It was many centuries before a device was invented capable of doing this work in which a draught animal provided the power.

By the year 2500 B.C. the Egyptian Pharaoh caused to be built the largest of all the pyramids, the Great Pyramid, the base of which is

110 Egyptian gardening: a scene from a tomb of about 2000 B.C.

Many roots, vegetables and shrubs other than the vine were now brought into cultivation, and the various rulers appear to have delighted in rearing exotic plants. In this scene the two men on the left are gathering fruit while the man on the extreme right is planting out seedlings. Two other men carry water from a canal or reservoir.

111 The impression of a Mesopotamian seal showing a shaduf, about 2000 B.C.

112 A shaduf depicted on an Egyptian tomb painting of about 1500 B.C.

113 The shaduf in use in Upper Egypt today

Perennial cultivation meant that plants had to be provided with water. Sometimes water was brought from the river or reservoirs by hand, but the first illustrations of a water-raising machine, the shaduf, belong to this period. A container was suspended from the end of a beam with a counter-weight. The shaduf obviously increased the volume of water that a single workman could raise.

112

113

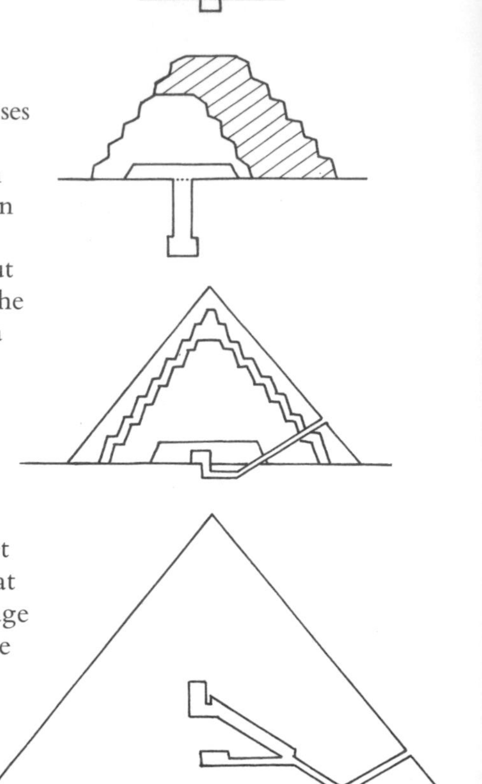

114 Diagram showing the evolution of the pyramid

In Egypt the period 3000–2000 B.C. was the great era of pyramid building. Their origin may be traced to the 'houses of the dead' (mastabas) in which, to prevent robbery, the dead were often buried in a chamber beneath the floor. In the 'Step Pyramid' (2800 B.C.) the mastaba was added to in two major stages. The pyramid of Seneferu at Meidum (2700 B.C.) still retained the mastaba and step structure, but was given a final smooth facing of limestone. Finally, in the apogee of all, the 'Great Pyramid' of Cheops, the mastaba and steps were eliminated, and the pyramid took on the familiar form.

115 The Giza group of pyramids

The 'Great Pyramid', seen here in the foreground, was a gigantic undertaking by any standards. Measuring 756 feet at the base, it rises to 482 feet and it has been estimated that its construction took 2,500,000 blocks of stone of an average weight of two and a half tons apiece. With a possible force of 100,000 men it may have taken twenty years to build.

756 feet square. It has been estimated that a total of not far short of two and a half million blocks of stone were required in its building and that the average weight of each block was some two and a half tons. It is extremely difficult to come to any reasonable estimate of what this means in terms of human labour, but more than one authority believes that a force of some 100,000 men, working during the season of flood – that is to say from the end of July to the end of October – were required to transport and lift the stone, and that this operation took in all some twenty years. Of course, at the quarry and at the pyramid there would have been permanent gangs of workmen quarrying and trimming the stone blocks throughout the whole of the year, but these gangs would not necessarily have been by comparison very large – perhaps a few thousand men each end. Such structures were, of course, gigantic undertakings demanding a considerable amount of planning, mathematics, practical geometry and the basic elements of engineering. But whatever the driving force that caused them to be built and however wasteful of man-power they may seem to us today the pyramids are only one lasting symbol of a period of great enterprise, for the Egyptians now were beginning to look much further afield for the raw materials they needed for their industries. Regular armed convoys of miners, for example, were being despatched into Sinai, there to quarry the copper ores, while regular expeditions were being mounted to go further and further up the River Nile in search of gold, ivory and other precious materials and even the timber required for their buildings.

Good timber, never plentiful in Egypt, was now nonexistent and any timber that could not be brought down the Nile had to be sought from other countries of the Eastern Mediterranean. Hence the Egyptians were obliged to build sea-going craft of wood. Their method of constructing these ships, however, betrayed their origin, for in fact they were little more than overgrown papyrus boats made of planking in which the edges of the planks were quite literally sewn together. They thus lacked either keel or ribs, and added strength was given athwartships by running a deck planking through mortises in the gunwale, while to prevent the vessel breaking its back a large stay was carried from stem to stern. Raised above the deck by forked timbers, this stay was maintained taut by a tourniquet passed through the rope and lashed to one of the uprights. Such a vessel could carry nothing in its hold and both passengers and cargo had to travel on the deck as, too, did the crew, who now appear to have abandoned the paddle and taken to the oar. One is left wondering whether or not the Egyptians borrowed the idea of rowing from other maritime nations of the Eastern

116 Egyptian wooden seagoing ship of about 2500 B.C.: a model reconstruction

Wooden seagoing ships were fitted with a bipod mast, but in order to prevent them breaking their backs at sea a rope was carried from stem to stern, held above the level of the deck by forked wooden posts. This rope was kept taut by a tourniquet. The crew, who now invariably rowed rather than paddled the ship, sat on stools on the deck. The cargo, too, had to be carried on deck since the hull was too frail to support the load.

117 A large rowed river-boat: a model from an Egyptian tomb of about 2000 B.C.

The overall deck and individual seats for the oarsmen can clearly be seen in the tomb-model. The steering-paddle was lashed to a vertical post, and passengers were carried on deck.

116

117

Mediterranean, and furthermore one must doubt the efficiency of the crews, for in their earliest form we find the oarsmen of these ships seated not on thwart-planks but on separate stools.

While in Egypt copper remained the normal tool-making metal, in Mesopotamia a subtle change was taking place in the field of metallurgy, for the smiths had found that a greater control of the product could be achieved by adding tin ore to metallic copper to produce their bronze, rather than by smelting the tin and copper ores together as they had done hitherto. Their products now became more and more standardized and the percentage of tin present in their bronzes stayed consistently around eight per cent – a very good figure for the production of a hard but non-brittle metal. It nevertheless remains something of a mystery as to where all the tinstone came from, and one suspects that the Mesopotamians were sending prospectors far afield looking for their raw materials. At this time, too, the Mesopotamians had discovered that many of the deposits of sulphide ores of lead, galena, also contained considerable quantities of silver, and they had devised a means of recovering the silver from this source. The galena was placed in a large roasting furnace, with the result that the lead was either volatilized or absorbed by a thick bed of ash while the silver, initially present as the sulphide, was reduced to the metal and could be recovered from the ash by washing. But it will be noticed that at this stage the lead was not utilized but allowed to go to waste.

Shortly after the year 2300 B.C. we find the whole of Mesopotamia unified under a single Akkadian ruler, Sargon the First, who pushed his frontiers up into Syria and thus ruled over a small empire extending from the Eastern Mediterranean to the Persian Gulf. As with his Egyptian counterparts there can be little doubt that Sargon's chief concern in this enterprise was to attempt to control the sources of supply of his raw materials and one is tempted, therefore, to suppose that much of the tin required for bronze-making came from the mountains of Syria and of Eastern Turkey, while Syria could also of course have provided timber and a trading outlet to the Eastern Mediterranean.

Thus, for a brief period we find the two most technologically advanced areas of the world ruled by two supreme monarchs. A scribe of a later period, however, records that one of Sargon's successors was forced to carry out a campaign to quell an uprising on his northern frontier. The attackers were a coalition of seventeen princes who ruled over Northern Syria and Eastern Turkey, and among these princes was one whose name betrays him to have been of Indo-European origin – that is to say, a member of a linguistic group whose homeland lay in a more northern area of Asia. If the scribe's account is true, then the Akkadians were already in contact with the people who were to have the most profound effect on the technology of the early Near East, for it was the Indo-Europeans who came from the steppes of Asia who, almost certainly, first domesticated the horse.

Before we move on to the next great epoch, however, we must say something more about both Egypt and Syria, for many small technological innovations appear during this period that show considerable sophistication.

The old glazed soapstone ornaments, the so-called Egyptian faience, were now replaced in both areas by a completely synthetic material. White sand was mixed with natron, a naturally occurring form of sodium carbonate, shaped and heated so that the whole mass fused. It was to this synthetic core that the blue glaze was applied. The first step in fact had been taken towards the making of true glass, for the fusion of quartz and soda with the addition of a little lime to make it stable, remains the basis of much of our glass even today. At this early period it is clear from Mesopotamian texts that the glass-makers were unaware of the need to add lime to ensure a stable glass. Fortunately for them, however, their raw materials already contained sufficient lime without the need for its deliberate addition. The synthetic core made of sand and soda, if overheated, would of course have become molten, and there exist many examples of this type of ornament in which the heating had been stopped only just before the object melted completely and lost its shape. So there is some justification in supposing that the discovery of glass resulted from observing accidents in which faience was heated until it melted. Just before 2000 B.C. we find in Mesopotamia the first appearance of true glasses. But the objects themselves betray the fact that the Mesopotamians had still not appreciated fully the potentialities of their new material, for instead of moulding the material while it was hot, as one might expect, it was worked in the cold and was largely cut and polished with abrasives in the same way that these early craftsmen dealt with harder decorative stones. Nevertheless it is evident that a great deal of experimentation was being carried on, for at this time we also find the appearance of a limited amount of lead in the glazes on faience objects. The effect of lead in a glaze, as in a glass, is to give it a greater brilliance, and while we have no idea how the discovery was made it is clear that the craftsmen who made faience were on the look-out for better ways of doing things.

118 A model of a plough-team from an Egyptian tomb of about 2000 B.C.

119 Part of the impression from a seal from Mesopotamia showing a ploughing scene, about 2000 B.C.

120 Symbols depicting ploughs from Mesopotamian seals (3000–2000 B.C.), and a reconstruction of this type of plough based on these pictures and on ploughs still in use in Iraq today

In Egypt ploughs remained virtually unaltered in design throughout this period, but in Mesopotamia a sole, as a separate piece of timber, was added, and later a seed-drill was inserted into the sole. In the seal from the very end of this period one can see the sower pouring the seed into the drill while ploughing is in progress, but the plough alone often appears as a symbol on much earlier seals.

The same spirit surely prompted the introduction of what we might reasonably call a gadget which was attached to the plough. Earlier we saw that the Mesopotamian plough started life very like the Egyptian one as a pronged stick dragged through the ground, to which stick later became attached a sole to make a more effective furrow. Now we find a vertical hole drilled through the forepart of the sole and a vertical tube inserted into this hole, the top of the tube being shaped like a funnel. This curious device was in fact a seed-drill and there exist a few illustrations of the sower pouring seed into the funnel. This gadget, which certainly ensured that all the seed ended up in the furrow, had a fairly long life, for nearly 1,500 years later we still find the same type of seed-drill attachment illustrated on the tiled walls of a royal palace in Syria, although by this time the drill may have become an archaism.

Prior to Sargon's conquests, the temple, although being the main building of any city in Mesopotamia, had not dominated the scene as it did from now onwards. Early temples had often been built on a low raised platform of mud-brick, but now it became the habit to build the temple on the top of a series of stepped platforms, so raising the temple far above the level of the remainder of the city. The building of these stepped platforms, or ziggurats, presented certain problems, for had they been constructed entirely of mud-brick they would have been unstable due to the sheer volume of the building, while it was beyond the economic resources of these people to have created ziggurats entirely of fired brick. Thus, in order to bind the whole structure together and prevent excessive movement, layers of reed matting were often incorporated between the layers of mud-brick, fired brick and stone being reserved only for facing.

The building of pyramids in Egypt and of ziggurats in Mesopotamia demanded not only some knowledge of geometry but also a uniform method of measurement. This, of course, is not meant to imply that uniform systems of measurement were devised simply in order that pyramids and ziggurats could be built. Indeed long before the Egyptians ever thought of building a pyramid they had needed to make an annual survey of agricultural land after each flood, and perhaps because of this the Egyptians became adept at surveying, while even as early as 3000 B.C. we find that standards of weight and of length were being maintained in temples and royal palaces in both Egypt and Mesopotamia.

Early units of length were based upon measurements that can be made from one point to another on the human body: thus throughout the Near East we find the cubit, from the point of the elbow to the tip of the middle finger; the span, from the tip of the little finger to the

121 The Ziggurat, Ur, during excavation

122 A reconstruction of Ur Ziggurat as it may have looked in antiquity, about 2000 B.C.

The ziggurat became the centre-piece of most Mesopotamian cities. It was essentially a stepped platform on top of which was built the temple. Ziggurats were built almost entirely of sun-dried brick, and in order to support the superimposed load of the building without the brickwork crumbling, layers of reed-matting were incorporated as the construction progressed. The matting served as a reinforcement in much the same way as do steel rods in a modern concrete building.

123 Diagram showing how Egyptian and Mesopotamian measures of length were based upon parts of the body

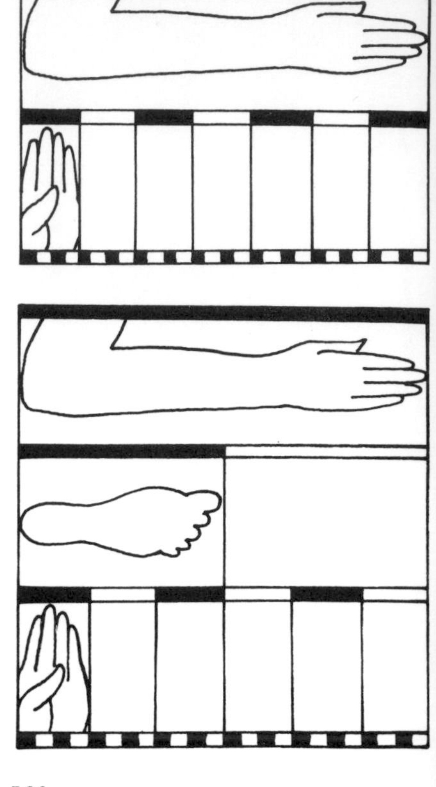

123

Elaborate buildings, and other construction work, demanded the use of standard units of linear measurement. The cubit, from the elbow to the tip of the middle finger, was in general use throughout the Near East. In Mesopotamia the cubit was further subdivided into feet, hands and finger-widths (inches) and was, hence, ancestral to our own system of measurement. In Egypt the cubit was divided into seven hands, each of four finger-widths. Needless to say the cubit was not of identical length in all regions.

124 Egyptian stone weights from Sinai, about 2000 B.C.

125 Egyptian balance as depicted on wall paintings of about 2000 B.C.

Trade in metal and other precious materials equally demanded standards of weight. These were all based upon the theoretical unit, the weight of a grain of wheat, and different people used various multiples of the grain as larger units. These varied very considerably from one country to another, so that a merchant trading in the Mediterranean, for example, would have to carry a set of weights appropriate to each country he visited. The Egyptian weights shown in this instance were those used to measure the dried fish ration of miners extracting copper ores in Sinai.

To judge by the accuracy of the weights themselves, balances must have been fairly sensitive. Simple two-pan balances were often depicted, but never in sufficient detail to show how they were pivoted or adjusted.

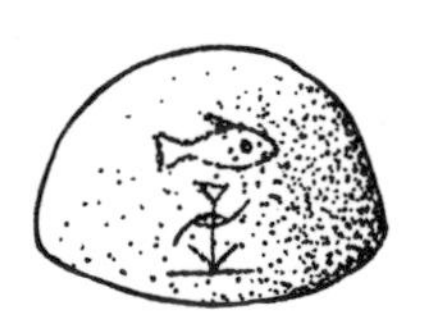

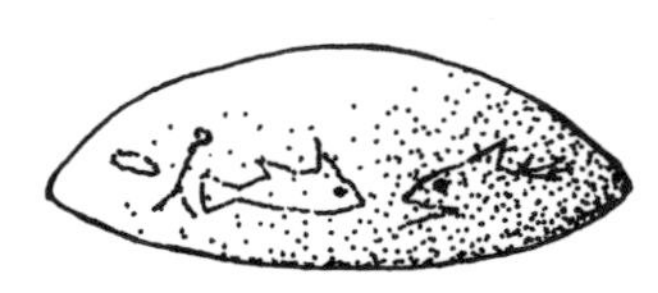
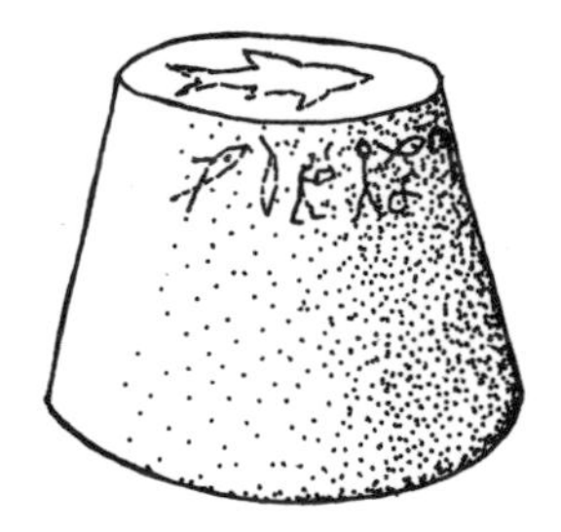

124

125

tip of the thumb of an outstretched hand; the palm, usually measured across the knuckle; the finger-width and the foot. Furthermore, the smaller units – spans, palms, finger-widths and feet – were generally related as subdivisions of the cubit. But very early in Mesopotamia and Egypt a totally different view developed as to how many subdivisons there should be to each cubit as well as what the subdivisions should be called. Thus the Egyptian royal cubit was divided into seven palms, each palm being reckoned as four finger-widths, giving a total of twenty-eight finger-widths to the cubit. Elsewhere in the Near East we commonly find the cubit divided into two feet, the feet into three palms and the three palms again subdivided into four finger-widths – a system that can be readily recognized as ancestral to our own measurement of feet and inches.

Since both Egypt and Mesopotamia had now become commercial countries in the strictest sense of the word, they had to have not only standard units of length but also standard units of weight in order to regularize their transactions. Initially balances were used only for weighing precious materials and the earliest units of weight were hence small, the major unit being the shekel, which was divided into a number of grains. Thus, while the grain itself was the hypothetical weight of a grain of corn, the shekel might vary from 120 grains to over 200. In time and as the need arose larger units – multiples of the shekel – were devised: the mina reckoned at anything from twenty-five to sixty shekels, and later the talent reckoned at sixty minas. While, however, the mina and talent were in use throughout most of the Near East, in Egypt a system of metric weights was in use, which seems to us today strangely ahead of its time.

The actual weights themselves were normally made of some hard polished stone with the denomination of the weight cut into it. In Mesopotamia, for reasons that have never been explained, these weights often took the form of a duck preening itself, but Egyptian weights tend to be more prosaic and are usually geometrically shaped blocks of stone with rounded corners and edges. As with measurements of length, units of weight have been found as standards in palaces and temples, but here again there is a very considerable variation in the value of the standards from area to area, and even from city to city. The balances were simple – equal armed scales with a pair of pans, and although we know of the balance as a symbol in early writing and from many illustrations, there is little to suggest at this period how and of what the fulcrum was made and, hence, it is very difficult to get an estimate of the accuracy of weighing possible with these early balances.

It is clear both from their achievements and from existing records that the Egyptians and the Mesopotamians, during the thousand years just reviewed in this chapter, had made considerable advances in the mathematical field. The Mesopotamians, for example, for no apparent practical purpose, had already learnt how to resolve simultaneous equations, while the Egyptians, through a study based partly on the annual flooding of the Nile and partly on the behaviour of the heavenly bodies, had established a reasonably accurate calendar. From their experience in surveying the land, moreover, the Egyptians had discovered how best to lay out a right-angled triangle, and indeed had found their own solution to the knotty problem of the square on the hypotenuse, using a system in which the hypotenuse was divided into the same number of subdivisions as the two equal sides of a right-angled triangle. This simple ruse allowed them to multiply and divide areas as well as lengths.

Few if any of the technologies so far considered require in their working any knowledge of mathematics at all, but in a somewhat roundabout way the ability to weigh accurately must have helped the metallurgist in more than just his commercial transactions. Thus in Mesopotamia we find an almost set formula for an operation performed by the goldsmith. The inscriptions usually read something like this: 'x minas of gold were put into the furnace and after heating y minas of gold remained; the loss as a result of heating equals x minus y minas'. Clearly the idea must have occurred to the Mesopotamian goldsmiths that gold could be obtained so pure that when put into the furnace it would suffer no further loss, and from this time on pure gold became the primary standard of exchange.

5

Chariots and Ships

(2000–1000 BC)

The policies of Egypt and Mesopotamia of pushing out their frontiers to include areas rich in mineral and other resources had their dangers. In modern parlance, apparently the administration in both areas was extended beyond its capabilities. This was serious enough, but the northern boundaries of Mesopotamia brought the people of the Euphrates Valley into contact with groups of tribesmen living in the hilly areas of Eastern Turkey and Northern Syria, and by 2000 B.C. an infiltration into this hill area of other people from more northerly regions of Asia had begun to take place. As we shall see later, some aspects of Mesopotamian civilization had already been diffused by way of the same hill areas into the Asiatic steppe and by this period the nomadic herdsmen had already acquired the wheeled vehicle, almost identical in its cumbersome shape and method of manufacture to the early carts of Mesopotamia, if indeed they were not its original inventors. But instead of the ox which was too slow or the onager which was not strong enough, the steppe-dwellers had apparently learnt how to tame and to harness the horse. The newcomers into the mountainous area north of Mesopotamia had thus on the one hand domesticated horses which they either brought with them or acquired from their northerly neighbours, while on the other hand from the Euphrates Valley they had learned sophisticated methods of working wood that so far had been applied only to the manufacture of such things as furniture. Within a very short space of time these hill-dwellers had discovered how to build a light manoeuvrable vehicle of war that could be drawn by a team of horses at great speed. The chariot was now to loom very large in the affairs of man.

This vehicle demanded a number of innovations. In the first place the horse could not be controlled either by a simple halter or by a ring through its nose; instead, the bridle and bit had to be used. The yoke and draught pole were left virtually unaltered, but the chassis and wheels

126 Faience tablet of about 1500 B.C. showing a chariot

127 Reconstruction of an early Egyptian chariot, based upon tomb paintings and on two surviving vehicles

128 Diagram showing how the yoke was adapted for use with horses, in its simplest form (above) and in the more elaborate Egyptian ogee-curved form of about 1500 B.C. (below)

The earliest true chariots were lightly constructed, highly manoeuvrable vehicles carrying a crew of a driver and one or two warriors. Their speed, relatively much greater than that of infantry, meant that military commanders were forced to adopt totally new tactics. The chariots were drawn by horses yoked in pairs and controlled by bridle and bit. Two Y-shaped pieces of wood were attached to the yoke, so as to fit the narrow shoulders of the horses. Early chariot wheels had only four spokes, which must have limited the terrain over which they could travel.

126

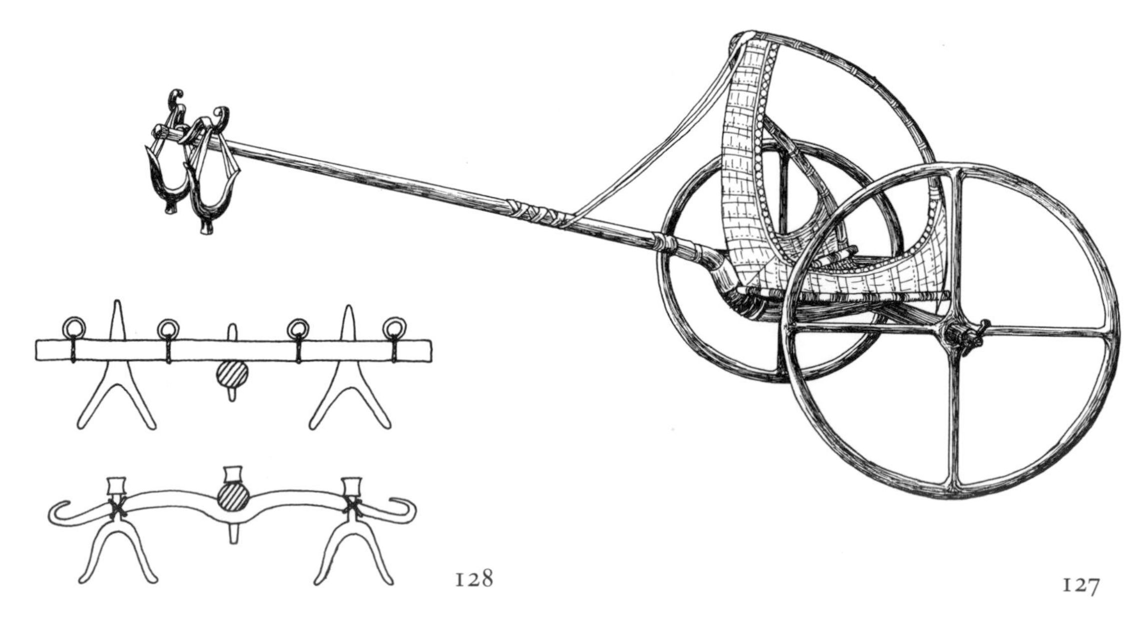

128

127

were completely redesigned. The chassis now appeared as a light wooden frame designed to do no more than carry the two or three warriors, while the wheels were made of a hub from which radiated four spokes set into a felloe of bent wood. When one considers the stresses and strains involved in driving a wheel hard over rough ground, one appreciates that hub, spokes and felloe would probably each have had to be made of a distinct type of timber, and that a wood suitable for one purpose might well not have served for the others. Although the only surviving chariots come from Egyptian tombs of a rather later date, it is equally clear that the wheelwrights were fully aware of the need to pick the right timber for the job: a hard, dense wood, little prone to cracking, for the hub; a straight, tough, inflexible timber for the spokes; and a straight-grained, easily bent timber for the felloe. Elm, oak and ash, still used in Europe today for the making of cart-wheels, were the same timbers as selected by the makers of chariots.

Obviously the chariot was a weapon that could and did revolutionize warfare. The speed of an army was no longer that of a marching column of infantrymen, although of course infantry remained the major part and backbone of any army. But the chariot could operate ahead of, and at a considerable distance from, the main force. Any army commander careless enough to allow his column to straggle ran the almost certain risk of seeing it decimated piecemeal by his opponent's chariot force. From now on the military commander had the means of carrying out the outflanking manoeuvre in the grandest possible style.

The political results of the introduction of chariotry are fairly self-evident. It need hardly be said that new regimes appeared both in Mesopotamia and ultimately in Egypt. The chariot, however, had more lasting effects and its introduction was to go a long way towards changing many basic technologies. All the evidence suggests, for example, that the people of the Mesopotamian Valley were now cut off from their main supplies of tin and that they began to seek other sources in the West by way of the Levant, thus bringing them even more into commercial contact with the people of the islands of the Eastern Mediterranean, that is to say Cyprus, Crete and the Aegean, as well as with people dwelling on the Anatolian coast.

The scrappy evidence that we have, largely representations on cylinder seals, a few clay models and sketches that can only be referred to as doodles, shows that by now at least two types of vessel were being used in the eastern end of the Mediterranean. The one was probably a broad-beamed vessel with a high sweeping bow and stern which at first sight appears to be simply an overgrown version of the reed boat.

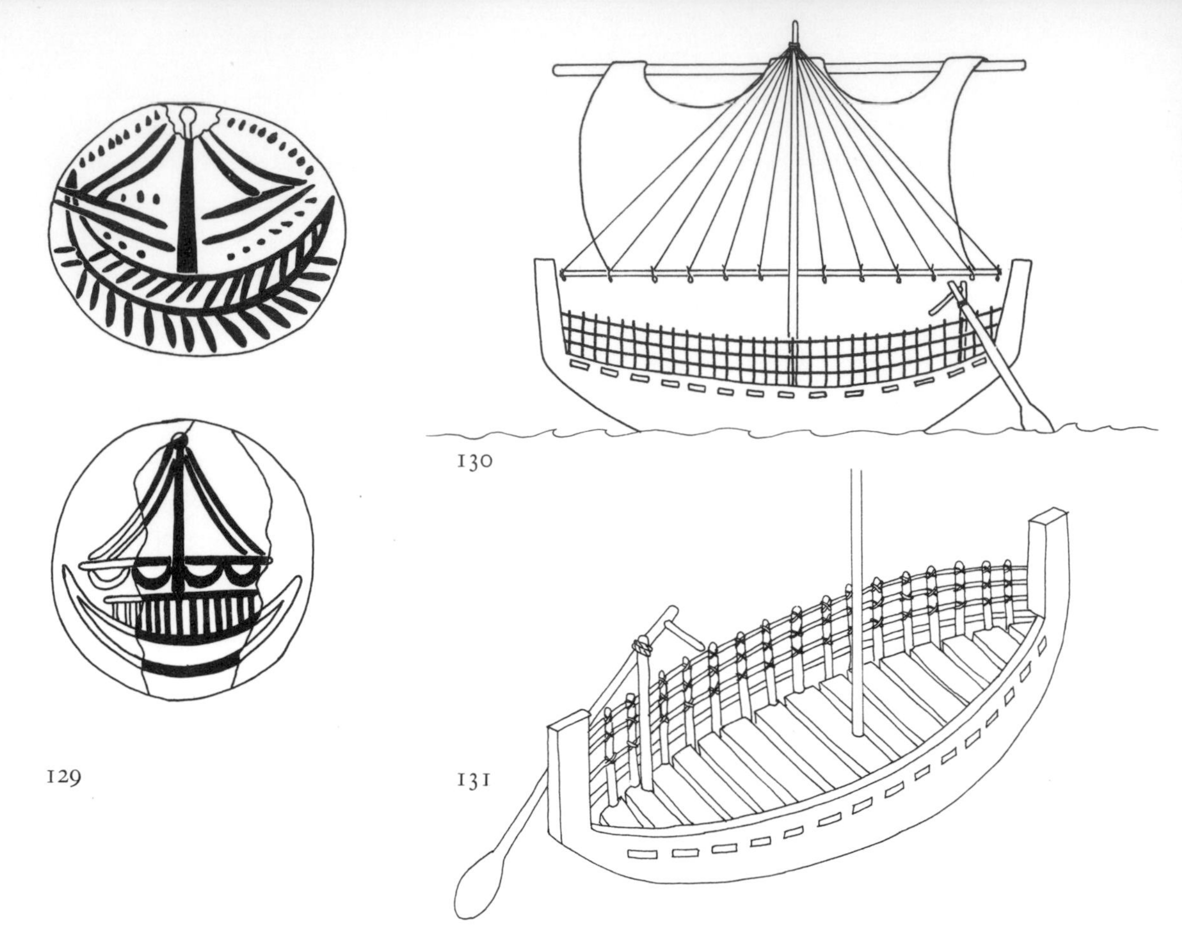
129
130
131

Even, however, if this were the ultimate source of inspiration the vessel must have been plank built. The second type of vessel, although equally high-sterned, was apparently less broad in the beam with a low bow rising vertically from a projecting 'ram'. This vessel looks as though it were essentially an extremely large tree trunk that had been hollowed and shaped and to which wash-boards and a high stern had been added. This is a type of construction which, without the high stern, can still be seen off the coast of Senegal today where it is used by fishermen who, it should be noted, are quite happy to sail the vessel despite its lack of keel.

Both types of craft were propelled by oars or by a square sail. Apparently the vessels were rowed by between five and ten oarsmen a side, although the smaller number is the more common, and on this basis we can guess that the ships ranged from perhaps thirty to fifty feet in length. Oars were, of course, essential when the ship was becalmed or when working into wind, but when the wind lay abaft a square sail was used, hoisted to a single mast stepped amidships, a practice very different from that used hitherto in Egypt. It is impossible to say whether these vessels carried a deck or not, but on the face of things it seems unlikely. Some representations would suggest that the wash-boards were attached to the hull by means of a series of vertical posts,

129 The impression of two seals from Crete, about 1500 B.C.

130 Reconstruction of a ship from the Eastern Mediterranean, based upon two badly damaged Egyptian tomb paintings of about 1500 B.C.

131 Conjectural reconstruction of this type of merchant ship

132 A Portuguese fishing-boat which possibly echoes some of the features of Mediterranean ships of this period

Merchant ships from Crete and the Levant that traded with Egypt were apparently short, broad-beamed vessels, rising at both stem and stern. Cretan seals give only the vaguest impression of what they were like, and the reconstruction is based very largely on badly defaced paintings from two Egyptian tombs. The vessels appear to have been decked at a level a little below the gunwale, and it was upon this that the cargo was carried, protected to some degree by hurdlework rising above the gunwale. The square sails differed from those of the Egyptians in that they were attached to the yard-arm only in the centre and at either extremity. It is far from clear whether the hull was strengthened by internal ribs, but the general form of the hull seems to have been preserved in small lighters still used in some Black Sea ports and in fishing-vessels used by the Portuguese.

132

and if this is so then the idea of internal, strengthening ribs was already present in embryo. The high stern to protect the helmsman, and also presumably the master, from the seas when there was a following wind is a feature that was to be retained in much Mediterranean shipping from this time on and is still to be seen in many local types of craft in use in that area today.

It is one thing to be able to say that the people of Crete and the Levant had vessels adequate for sailing on the Mediterranean: it is quite a different matter to say with any certainty precisely where they went and in what commodities they were dealing. However, even as early as 2500 B.C. it is clear that a number of small colonies of people coming either from the eastern end of the Mediterranean, or more probably from Malta or Sicily, were established in Southern Spain and Portugal. These people, living in small fortified cities and using bronze for tool-making, were surrounded by an indigenous population who still relied on stone tools and weapons. The area is rich both in copper and in tin so that one suspects that here at least was one source from which the people of the Eastern Mediterranean were deriving the metals so essential to their technology.

Among the countries to profit from the presence of these early merchant venturers was Egypt, where we now find that bronze rather than copper had become the general metal for tool-making. From this period, too, we begin to find small blue beads made of Egyptian faience spread widely amongst the scattered and backward people of Western Europe, which allows us to think that quite a considerable trade had been built up down the length of the Mediterranean.

Insofar as we can determine, by 2000 B.C. tinstone was still being added to metallic copper whenever it was required to make bronze, so that it is highly likely that what was being imported into the eastern end of the Mediterranean was either the tinstone itself or more probably the already formed alloy. This does not mean, nevertheless, that no further progress had been made in the Eastern Mediterranean in the field of metallurgy. During the next five hundred years the whole process of bronze-making became extremely sophisticated. Shortly after 2000 B.C. we know for certain that the bellows had been invented and were being used as a means of raising the furnace temperature, for we find an inscription on one of the Mesopotamian clay tablets which is a demand for the skins of two large he-goats for the making of bellows for a bronze-founder: while at a slightly later date an Egyptian tomb painting shows us twin bellows in use being worked by foot. The bellows must have allowed a far larger scale of production than

133 Drawing from part of an Egyptian tomb painting of about 1500 B.C., showing foundrymen casting bronze doors

134 Diagram showing how the crucible was probably handled

135 Drum-bellows in use until quite recently in Rhodesia

With the introduction of the bellows for raising the temperature of the furnaces, it became possible to heat larger volumes of bronze than was possible when the fire had to be blown by lung. The bellows were either whole skins or skin-covered drums worked in pairs. In this scene from an Egyptian tomb the men working the bellows must be envisaged as rocking from side to side as they tread down first on one drum to expel the air, and then on the other. The function of the string attached to the skin of the bellows is to raise the skin, so filling the drum for the next tread. Similar hand-operated bellows were still in use by blacksmiths in Rhodesia until quite recently.

In the Egyptian scene the foundrymen are casting a bronze door, and the metal is being poured into a series of cups leading into the large clay mould. A previously cast door can be seen above and to the right of the foundrymen.

133

135

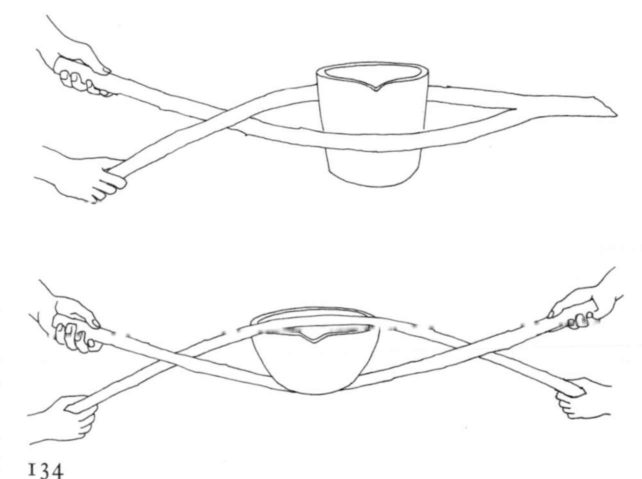

134

did the blow-pipe, permitting not only the production of more tools and weapons but also, because greater quantities of the metal could be melted at the same time, the making of larger castings. A tomb painting in Egypt illustrating the casting of a large bronze door shows us that the mould had a whole series of sprue cups through which the molten metal was poured, a situation that would have been quite impossible if all the metal had had to be made fluid by the older expedient of raising the furnace temperature by means of mouth-blown pipes.

Chemical analyses of bronze implements show us that up to this point metallurgists had had to be highly selective in the copper ores they used. They were unable to use ores containing sulphur because it formed a gas within the molten metal which, on cooling, became porous and unserviceable. It was now discovered, however, that by roasting the sulphide ores in an open fire the worst of the sulphur could be eliminated and that the ores could then be reduced in a normal furnace to provide a sound metal. Since sulphide ores are, in fact, of far more common occurrence than other types of copper ore, this discovery made a much larger quantity of metal immediately available.

Finally, the old practice of adding tinstone to metallic copper gave way to a different method of alloying. The tinstone was first reduced to metallic tin in a furnace and the two metals were then heated together to form bronze. This system allowed complete control over the relative proportions of the two metals being used, and from now on bronzes with low and with high tin contents were being used for specific purposes. Thus for the making of mirrors a very high proportion of tin was added, which while making the metal far whiter also made it very brittle and thus useless for tools. For tools and weapons a proportion of about eight per cent of tin was retained, but where for example a bone handle had to be fitted to a bronze knife the rivets were often made of an alloy containing even less tin so that they would be more easily hammered without becoming brittle, since they could not be annealed without damaging the bone handle.

Bronze now became so common that it was in everyday domestic use throughout the whole of the Near East, and stone tools of any kind became increasingly rare. Nevertheless, despite the large quantity of bronze available, from about 2000 B.C. onwards iron tools, weapons and even ornaments of iron were occasionally being produced. Iron does not normally occur in nature as the metal, but usually as an ore, although metallic iron is to be found in very small quantities in some meteorites. Thus while some of the very earliest iron objects may be seen as having been manufactured from meteoric iron it is unlikely that this applied to most of them, and it is fairly evident that over a long period, perhaps some seven hundred years, metallurgists were experimenting with iron, never quite managing to master the techniques required in its production.

Unlike any of the metals that were so far known – gold, silver, copper and tin – iron could not be made fluid in the furnaces being used by these early metallurgists, for the temperature was always too low. Even so, if the conditions were right, the ores of iron could be reduced to the metal, but the end-product as it was removed from the furnace was a hard spongy mass that looked very little like a serviceable metal. To make anything of it, it had to be repeatedly reheated and hammered while red-hot in order to forge it into a bar of solid metal. This process was quite unlike any metallurgical technique so far known.

Unfortunately, we know remarkably little about the discovery of iron-working, and although it is easy to suggest that in the first instance a very pure red oxide ore, red ochre, was reduced in small quantities and was heated and hammered as one might hammer and anneal bronze, we still know next to nothing about how and where the full-scale production of iron really started. Historically, the first people that we know of who had a limited but steady supply of iron were the Hittites, whose kingdom occupied very approximately the eastern half of modern Turkey. An existing letter from a Hittite ruler to the Egyptian Pharaoh tactfully explains that he cannot let him have iron swords at the moment as good iron is not to be had. In short, we are not even certain that the Hittites manufactured their own iron, and they may well have acquired it from a neighbour. Many authors today suggest that large-scale iron-working may have started within a region roughly defined by the sides of a triangle running from the south of the Caspian Sea to North Syria to the south-eastern end of the Black Sea, but concrete evidence of this is still lacking.

Although the ores of iron are far more widespread than those of copper, for indeed there are very few countries that can produce no iron ores at all, the technique of winning the metal was so different from that of copper production that the spread of the knowledge of iron-working appears to have taken a very considerable time. Not only did the craftsman have to learn to forge his metal once it had come from the furnace, but he also had to learn to shape it while still red-hot and to weld one piece of metal to another by heating them both to red heat and then hammering them until they welded. Furthermore, it was necessary to exercise a far greater control over the atmospheric conditions in the furnace than was necessary during the smelting of copper, for the iron ores, never becoming molten, had to be converted to the metal by carbon monoxide gas in the furnace, which was not essentially the case with copper ores. It would have been impossible to produce any quantity of iron at all without bellows, or some other system of creating a steady and controlled forced draught, and although in Mesopotamia the bellows first appear, as we have seen, as a bronze-founder's tool it is far from certain that the idea was not borrowed from people of the North already working small quantities of iron.

In order to work the red-hot metal, tongs, a heavy iron hammer with a long handle and a substantial anvil were essential, whereas all the hammering processes with copper and bronze could have been, and in fact were, carried out in the cold, thus avoiding the need for forging equipment.

It has been suggested that iron-working remained the monopoly of the Hittites and their immediate neighbours over a long period of time, and that it was only with the fall of the Hittite Empire that iron-working spread to other parts of the Near East. Even if this hypothesis is correct, the metal was used only for the making of weapons, and although from 1500 B.C. onwards iron weapons became increasingly more common it was not until the last century before the year 1000 B.C. that we find the metal in use throughout the larger part of the Near East, and even then it was still not being produced in Egypt. As we shall see, shortly after the year 1000 B.C. the technology of iron-working spread more rapidly and then, and only then, do we begin to find tools rather than weapons made of this metal. Thus for the whole of the period being considered in this chapter we must remember that bronze was the universal metal for the making of all tools.

Apart from the improved techniques in working tin and copper, from the year 2000 B.C. onwards in both Egypt and Mesopotamia the making of glass became increasingly more sophisticated, a large part of the production going into the making of glass vessels and beads. Small glass bottles were usually made either by dipping a friable core of sand and some organic adhesive into a crucible of molten glass, or less probably by covering the surface of the core with broken and finely ground glass material and then fusing it in an oven. In both cases the core was extracted at the end of the operation. Mesopotamian texts make it quite clear that they had mastered the technique of producing quite a wide range of colours in their glasses. Copper ores were used to give turquoise-blues, cobalt ores to produce darker blues, iron ores to provide yellows, and tinstone to make white, opaque glass. Threads of differently coloured glass were often arranged in intricate patterns

136 Egyptian glass vessel of about 1500 B.C.

Glass now began to be shaped in the hot, plastic state rather than being worked cold by abrasion. Glass vessels were made either by dipping a sandy clay core into crucibles of molten glass or by winding threads of glass around a core, later reheating and rolling the surface to make the threads fuse and to flatten them. The surfaces of the vessels were often decorated by applying threads of differently coloured glass which were, again, reheated and rolled into the surface. The same decorative technique was used in the making of beads and other small ornaments.

on the surface of the object that was being made, which was then, while still hot and plastic, rolled gently over a flat surface, so welding the pattern of threads into it.

We saw earlier how silver was often smelted from lead ores, with the apparent wastage of the lead. Indeed, lead as a metal had been known from perhaps 3000 B.C. for it is comparatively easy to smelt, but little use had been found for it since it was far too soft for the making of tools and too unattractive for the making of personal ornaments. From about 1500 B.C. onwards, however, we begin to find lead being used not as a metal but as an ingredient in both glasses and in bronzes. Lead, if present in a sufficiently large quantity, alters the characteristic behaviour of a glass on cooling. Glasses, for example, made only of an alkali, such as potash, and silica contract very considerably on cooling, so that if one attempts to use such a glass to cover the surface of, say, a pot or a brick, one finds that on cooling it contracts more than the material to which it has been applied and as a result it cracks. But by the addition of large quantities of lead this shrinkage is greatly reduced, allowing one to apply a glaze to an earthenware surface. This fact was evidently grasped by the Mesopotamians shortly before the year 1000 B.C., when we find early experiments in the glazing of brick and tiles, although for some unexplained reason little attempt seems to have been made to glaze pottery at this time.

137

138

For the bronze-founder, the introduction of lead into his alloy in small quantities allowed him to achieve what had hitherto only been possible with a great deal of very careful forethought and preparation. The addition of between five and ten per cent of lead to bronze does not greatly alter its properties as a metal for the making of tools and weapons, but it does alter its properties when molten, for the alloy becomes far less viscous at low temperatures and thus easier to pour into an intricate mould. We know by examining earlier castings that ordinary tin bronze often cooled so rapidly that all the impressions of a mould were not taken up by the metal, leaving flaws in the casting that were either ignored or had later to be filled. By adding lead to the bronze this difficulty was largely overcome and as a result not only did castings now become far more elaborate – indeed even weapons verged upon the baroque – but also new mould-making techniques now became very popular.

To achieve really elaborate castings the object, perhaps either a small statue or the handle of a sword, was first modelled in wax. The wax was then covered with a layer of fine clay and after this was added a thick outer coating of coarse clay. The mould was allowed to dry and then heated so that the molten wax could be poured off while the coatings of clay became fired like pottery. The metal could then be poured into the hollow left after the removal of the wax so that the ultimate casting took on precisely the shape of the wax model although, of course, the mould had to be broken open to get at the casting, and thus could not be used a second time. The technique was also adapted to the making of small statues of bronze which, had they been of solid metal, would have been so extravagant as to be impossible. First a core of clay rather smaller than the intended statue was made and allowed to dry, and to the surface of this was applied a layer of wax the thickness of which would be the same as that of the metal in the

137 Lead-glazed jug from Mesopotamia, about 800 B.C.

138 The lead-glazed brick façade of a gateway, Babylon, about 800 B.C.

Lead as a metal alone was apparently too soft and unattractive to have found many uses. The metal was, however, employed as an ingredient of bronze, glass and glazes. In the latter case, glazes began to be applied, in Mesopotamia, to bricks and tiles, often with additional colouring materials such as the salts of copper. Although of a slightly later date, this glazed brick façade gives a fair impression of the use to which such bricks were put and of the effect they could produce.

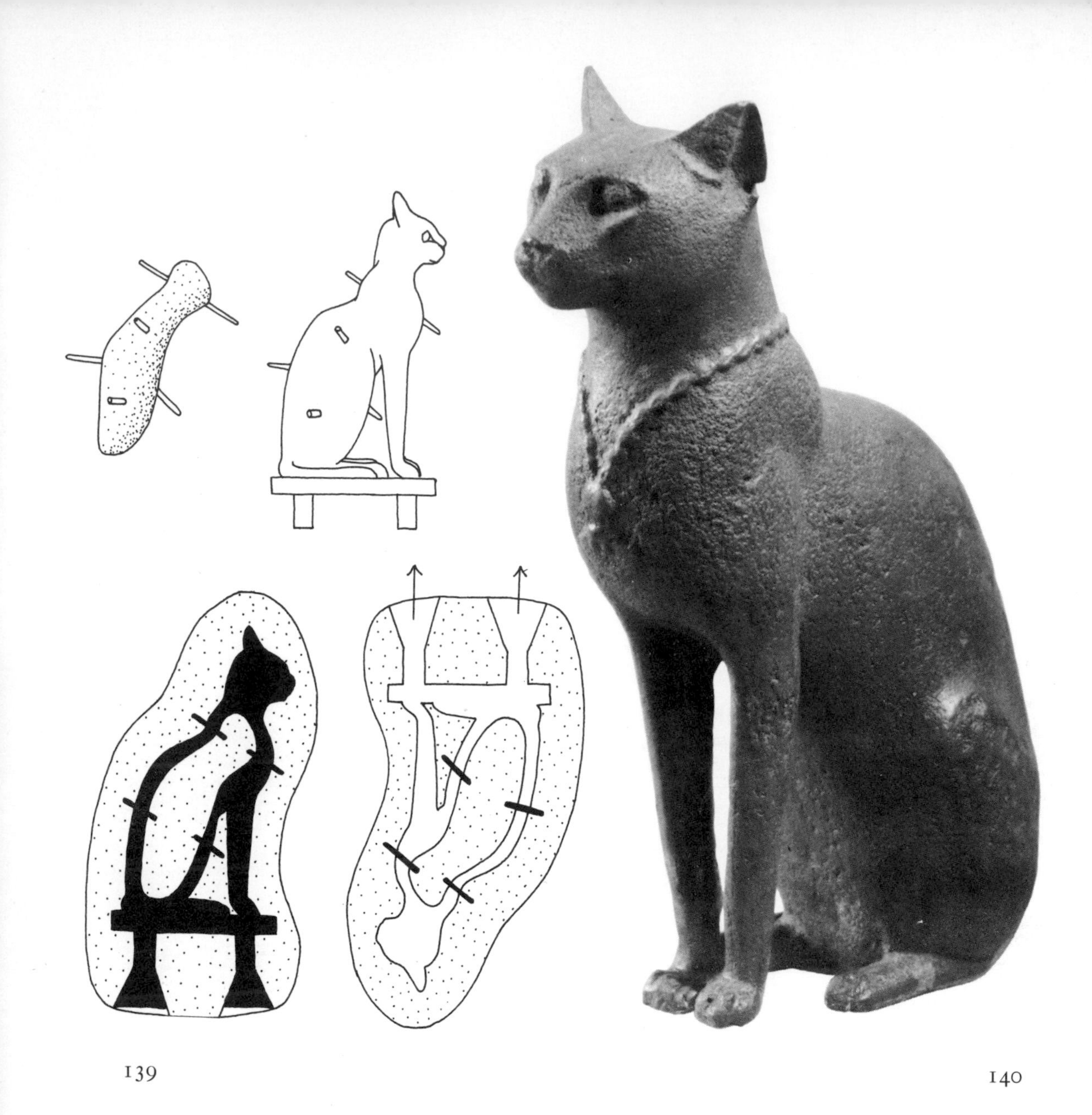

139 140

139 Diagram showing the stages by which Egyptian foundrymen cast a bronze statuette

140 Bronze statuette of a cat from Egypt, about 1500 B.C.

With improved furnaces also came better methods of casting. Objects to be cast were often modelled in wax. The models were then covered with clay to form the mould which was heated to remove the wax and fire the clay. The space left after the wax had been poured off was filled with molten metal and the mould was broken open to recover the casting. By incorporating a core of clay within the wax model it was possible to produce a hollow casting, so as to economize on metal. The general principles of 'lost wax' casting is still used today in fine art foundrywork.

finished statue and into which was modelled all the details. The core and wax were then coated with clay, dried and heated to remove the wax, and molten bronze was poured into the space left between the core and outer mould. Later the core might be removed although this was not invariably the case. A life-size statue could have been made in this way using eight or more pieces which were later joined. The result was far superior to the older technique of covering a wooden sub-structure with sheets of metal that had to be pinned in place. This technique of lost-wax casting in fact was not very different from that used by the bronze-founder today, although really large single castings remained beyond the early craftsman's ability.

Throughout this period of a thousand years many technologies remained virtually unaltered. Quarrying, masonry, the making of pottery and bricks and most of the techniques of agriculture saw little change. Apart from the technologies already described most of the improvements that took place lay in the domestic field, in such things as weaving and the production of household luxuries.

It will be remembered that the earliest looms that we know of were simple frames laid out horizontally at a short distance above the ground. In countries where it was practical to weave in the open air through most of the year this type of loom was perfectly satisfactory, but probably as the result of contact with more northerly people this period saw the introduction of a loom in which the frame was set vertically and thus could be housed indoors. The hanging warp threads had to be held under tension, and this was achieved by attaching a clay or stone weight to the lower end of the bundles of threads. Alternatively the warp threads may have been looped continuously round an upper and a lower horizontal beam so that the length of cloth that could be produced was in fact just a little less than twice the distance between the two beams. This arrangement was apparently almost unique to Egypt, for elsewhere in the Near East the warp-weighted loom now became the universal piece of weaving equipment.

Another device that we may almost certainly attribute to people who lived to the north of Mesopotamia was the plough with a proper share and sole. Hitherto the plough even in Mesopotamia, where it was fitted with a small sole, had done little better than make a scratch in the surface of the soil. The plough with a share and flat sole was designed to dig far deeper into the soil and thus make a better furrow into which to plant seed. With the relatively light soils of Egypt and the Mesopotamian Valley the older type of plough was perfectly satisfactory even if much of the seed was shallowly planted, but in the

141 Reconstructed drawing made from a damaged tomb painting showing a vertical loom. Egypt, about 1500 B.C.

142 Vertical loom with weights holding the warp threads under tension as depicted on a Greek vase of about 500 B.C.

143 Simple vertical loom still in use today, from Jordania

The older form of loom in which the warp threads were stretched at a short distance above the ground appears to have been largely replaced in the Near East by a loom in which the warp threads hung vertically from an upper beam. In the Egyptian wall painting the threads seem to have been held under tension by a lower beam around which they were looped. In all probability such looms varied little from those still in use in Northern Nigeria today.

Elsewhere, the warp threads were held under tension by weights attached to their lower ends. Although the weights made of stone or clay have often been found, there are no contemporary pictures of this type of loom, and to get an impression of what it was like we have to look at later illustrations on Greek vases. The vertical loom had the advantage of occupying less ground space than the horizontal loom and could thus be more easily housed indoors.

141

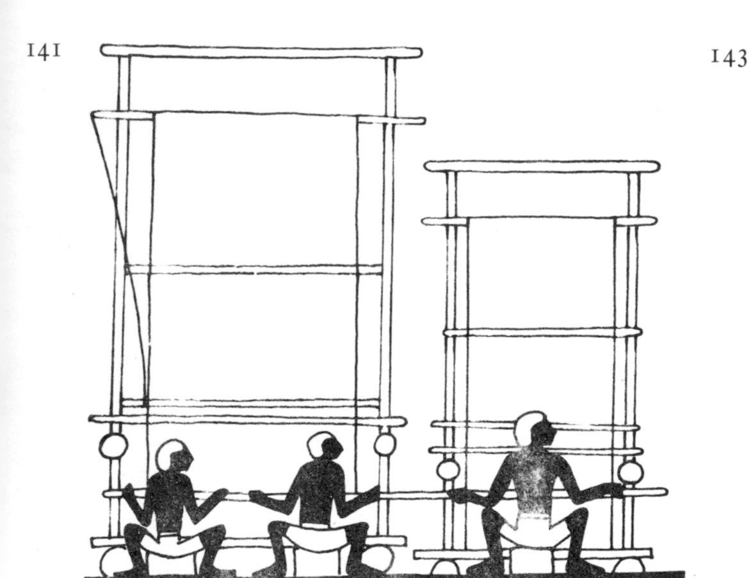

143

142

144 Drawing made from the impression of a Mesopotamian seal of about 1500 B.C., showing a plough with sole and single handle

145 Drawing of part of the decoration on a Greek vase of about 600 B.C., showing a plough with sole and single handle

146 Reconstruction of this type of plough based upon a number of decorated Greek vases and upon ploughs still used in Anatolia today

In Mesopotamia, but not in Egypt, the two-handled plough was replaced by one with a single handle and a far larger sole. This type of implement had the double advantage of making a rather wider and deeper furrow, while leaving the ploughman a hand free with which to goad his ox-team. He could, therefore, plough alone without an assistant to do the goading. In subsequent centuries this was to become the most common form of plough in Greece and other Mediterranean countries, while in only slightly modified form it is still to be seen in use today.

more northerly countries where the soils were heavier and the time required for germination of the seed was longer this type of plough could have been disastrous, and so we assume that it was in these areas that this much heavier piece of equipment was devised. It appeared in Mesopotamia considerably before 1000 B.C., but in Egypt it was not adopted until nearly a thousand years later.

Even by 2500 B.C. we have some evidence that in Egypt at least oils were being pressed from various fruits. The machine used for this purpose was a simple bag press in which the fruit was placed in a cylindrical cloth bag supported at each end by upright posts, which was then twisted by two operators using simple bar handles. Although presses of this sort were undoubtedly used for extracting the oil from olives it is equally clear that they were being used to extract aromatic materials from other plants to be used as flavourings and as scents. Indeed, during this period there grew up a considerable industry and trade in cosmetics, for which purpose many flowers and herbs were deliberately cultivated. Merchant seamen from Egypt were certainly exploring parts of the east coast of Africa and the upper reaches of the Nile in search of exotic materials for this trade, while from the Mesopotamian delta it seems more than likely that merchantmen were occasionally travelling as far as India in search of spices.

These early cosmetics were very largely unguents made of animal or plant fats and oils, scented with a multitude of aromatic materials. Indeed the unguent bottle, either of stone, glass or clay, now became a familiar household object amongst the more wealthy citizens of the Near East. But it was apparently in Cyprus that there developed an industry based upon the extract of the opium poppy, laudanum. In many parts of the Near East have been found small hand-made bottles shaped like the dried head of an opium poppy. These jars bear a peculiar decoration which seems to imitate the scar on the face of the poppy which was made in order to extract the opium-bearing sap. To the wealthier members of society this drug was in all probability used much as we would use aspirin today: to relieve headaches and hangovers and even, perhaps, to keep the baby quiet. In any event it was sufficiently treasured to allow the bottles to be included in many graves in case opium should be required in the afterlife.

In the last few centuries immediately preceding the year 1000 B.C. the whole of the Near East underwent a period of considerable turmoil due to the intrusion of people from Central Europe and Northern Asia, and many writers have seen the spread of the knowledge of iron-working as a part of this breakdown of the established order in the Near East.

147 Bottles in the form of dried poppy-heads from Cyprus, about 1500 B.C.

Juices were extracted from a wide variety of flowers, herbs and spices and compounded with fats and oils to provide unguents. Unguent bottles are a common feature of the homes and graves of the wealthy although the precise nature of the great majority of these cosmetics and balms is quite unknown. The shape of one type of bottle, however, does betray its probable contents. Made in Cyprus, and closely copying the form of the dried head of an opium poppy, it is found widely throughout other countries of the Near East.

While Greece, Anatolia, Persia, Mesopotamia and Syria were largely overrun by invaders, the Egyptians had to contend with an attack by both land and sea by people who appear to have been dispossessed by the northern invaders. The Egyptians called these intruders, whom they managed to repulse, the People of the Sea and they have left us one excellent record of a naval engagement that took place. This relief from the temple of Rameses III allows us, at last, to say something rather more concrete than we have so far been able about the design and function of shipping in the Eastern Mediterranean during this period. The Sea People are depicted as manning boats high in the bow and stern, the stem and stern-posts incidentally surmounted by animals' heads. By contrast the Egyptians are seen using vessels in which the keel has been extended to form a ram, and in which the mast is now set amidships. In general appearance the Egyptian ships would seem to be very similar to those vessels already in use by Egypt's more northerly neighbours before 2000 B.C., and it is a matter of question as to whether the Egyptians either manned or built their own navy, while some authors have suggested that much of the Egyptian naval enterprise during these centuries depended upon mercenary sailors from the Levant.

148

149

148 Part of the decoration of an Egyptian tomb of about 1500 B.C., showing merchant ships in port

149 A modern model reconstruction of this type of ship

Despite her very considerable overseas trade, Egypt adhered closely to ships whose design still echoed that of the reed-boat. Although about seventy feet overall, ships were still apparently without keels, and support was still given by a rope carried from stem to stern, as in the case of ships a millennium earlier (Figure 116). The cargo was carried on deck. A single mast was now stepped amidships, and the sails were furled by lowering the yard-arm. The cargo in this instance includes elephant tusks, trees with their roots contained in baskets, and monkeys.

150 Part of an Egyptian tomb decoration of about 1200 B.C., showing the Egyptians repulsing invaders

151 Part of the decoration on a vase from Greece, about 1200 B.C., with missing areas restored

152 A reconstruction of how the ships of the Egyptian tomb decoration more probably looked

Of a rather later date is the picture of the Egyptian navy repulsing invaders from the Eastern Mediterranean. The Egyptian artist, probably more familiar with reed-built boats, appears to have distorted the shape of both types of ship by giving the keels an excessive curve. The Egyptian ships were more likely to have been long, low vessels with a ram, the rowers protected by wash-boards or awnings. They may even have been ships manned and designed by mercenaries, for they show a strong resemblance to vessels at that time in use in Crete and Greece. The invaders' ships seem to have differed only in detail from the earlier merchant ships from Crete (Figure 130). The Egyptians are seen using grappling-irons, while both types of ship are fitted with fighting-tops at the mast-head – a feature seldom to be seen in Mediterranean shipping.

150

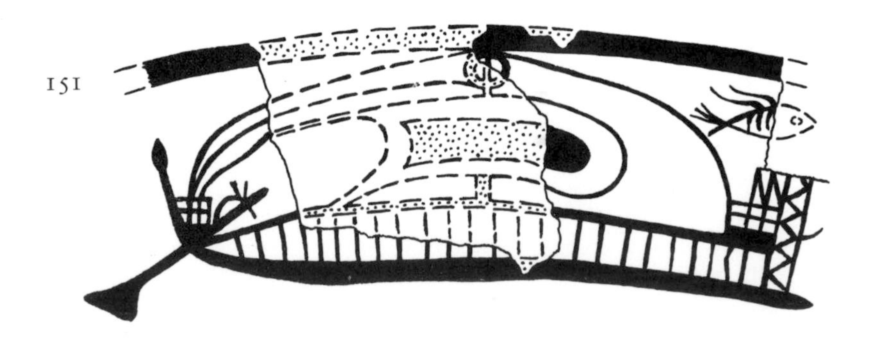

151

152

Whether this was so or not, these tomb decorations show certain innovations. For example, the ships of both the Egyptians and their opponents are fitted with fighting tops to the masts, from which an Egyptian, in one instance, is clearly seen slinging a shot. Furthermore, the Egyptians were using grappling-irons thrown from the end of long ropes as a means of hauling ships closer together either to board them or perhaps as a means of bringing the ram to bear on the enemy's ship. The ships of both the antagonists are fitted with wash-boards above the gunwale and screens in the fore and after peaks to protect the crews from the seas.

The Egyptians, however, were perfectly familiar with Mediterranean merchant shipping and from a tomb of a slightly earlier date than the temple of Rameses we have what is probably an Egyptianized view of these broad-beamed ships with a high bow and stern; vessels which appear to be only partly decked and on which may be seen some of the cargo contained in large jars, while above the gunwale runs a rather crude-looking railing

It is difficult from tomb paintings to determine how big a ship is supposed to be: since as a convention the skipper is often depicted far larger than other members of the crew, the figures are not a very useful guide to scale. However, in recent years skin-divers have discovered a wreck off the southern coast of Anatolia which goes some way towards helping us reconstruct this kind of vessel. In this case the

ship had been carrying ingots of copper which apparently had been laid directly into the hull of the ship although separated from the timbers themselves by a substantial layer of brushwood. The ingots were fairly heavy and one suspects that the brushwood was put there simply to protect the hull from damage should the cargo move. Although the ship has very largely disintegrated, the spread of cargo on the sea-bed suggests a vessel of some thirty or forty feet in length. For want of more reliable data, we must suppose that this would have been about the average-sized cargo vessel of this period, while this figure is to a degree supported by the fact that the ships shown in the temple of Rameses were rowed by seven oars on either side.

The ingots of copper recovered from this wreck are of a type hitherto occasionally discovered during normal archaeological excavation as well as being depicted in a few Egyptian murals, while their shape was used in Crete as a pictogram on records made on clay tablets. Each was a rather flat rectangle of metal some two feet long by one foot wide with four protuberances at the corners. Their shape is vaguely like that of the skin of an animal without head or tail, which has resulted in them being referred to as ox-hide ingots, although the protuberances may have been added to the castings simply to make them more readily handled. Nevertheless, the ingots clearly represented real wealth and may have served as an early form of currency.

By the end of the period under discussion in this chapter shipping had evidently become an important means of transport and communication in the Near East, although during the thousand years under review, due perhaps to a lack of information, we cannot honestly say that there were many vast or great changes in ship design. The chariot

153 Copper ingots in the form of an ox-hide from a wreck off the south coast of Turkey, about 1200 B.C.

Copper was traded in the form of flat ingots of metal often provided with four protruding 'handles', presumably to make handling simpler. A number of individual ingots of this type have been discovered, but those illustrated here came from a large group resulting from the under-water excavation of a wreck off the south coast of Turkey.

Ingots of this type are also often depicted in tomb paintings (see the man carrying an ingot, Figure 133, right) and even as an ideogram on Cretan clay records. In their wide geographic distribution and their comparatively uniform shape may be seen a step towards the idea of currency.

by contrast is more often depicted, and actual examples have been recovered from Egyptian tombs. During this period a number of changes took place in chariot design. Early chariots had an axle centred to the platform upon which the charioteer stood, so that he stood directly over the axle, while the wheels were invariably four-spoked. Somewhere around the year 1300 B.C., however, two changes were made. First, the number of spokes was increased to six, which would suggest that a greater loading was being put on the wheels, and secondly the axle was moved to the trailing edge of the platform. The charioteer was thus no longer balanced over the axle itself but his weight was carried partly by the wheels and partly by the yoke over the shoulders of the inner pair of horses. On the whole this seems a far more practicable arrangement than that in the earlier form of chariot, for the weight of the charioteer could now be used to bear down upon the yoke and thus prevent it riding up to force the throat-band against the windpipe of the inner pair of horses. The yoke, furthermore, was redesigned, for in its original form it was more suitable for the broad shoulders of an ox than for those of a horse. Into either end of the yoke was now mortised a new member in the shape of an inverted letter Y. The branches of this fork bore down on the shoulders of the horse, which were protected against rubbing by the addition of a small pad. Who was responsible for these changes is hard to say, but yokes of the kind just described have been found in Southern Russia and in China, while in the latter area many more than six spokes had been added to the chariot-wheel well before 1000 B.C. On balance, one suspects that the ideas were developed and diffused from the people of the Asiatic steppes.

154

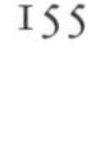

155

156

154 An Egyptian tomb painting of about 1200 B.C. showing a chariot with wheels with six spokes

155 A Hittite relief of about 1200 B.C. from Southern Turkey

156 Reconstruction of this type of chariot

157 Diagram showing how, by shifting the axle to the rear of the chariot, the weight of the crew was carried to a greater degree by the yoke

Towards the end of this period, in the twelfth century B.C., changes were made in the design of chariots. The wheels, which hitherto had only four spokes, were given six spokes, while the axle was moved from a position in the centre of the cockpit platform to its rear edge. Both changes were seemingly designed to allow the vehicle to move over rather rougher ground: the added spokes prevented the wheel from buckling, and the change in position of the axle meant that the weight of the crew was partly carried by the yoke bearing down on the horses' shoulders, thus correcting the see-sawing effect when moving over difficult terrain.

158 A relief from Southern Turkey of about 1000 B.C. showing a mounted warrior: one of the earliest representations of a horse being ridden

The supremacy of the chariot as a mobile weapon was soon to be challenged by cavalry. Enough had been learned about the control and training of horses to allow mounted men to fight from their backs. Early pictures suggest the use of a saddle-cloth or even a rudimentary saddle, but stirrups seem not to have been thought of until another thousand years when they began to be used by nomadic Asiatic people. Their more general adoption began with the invasions of nomads at the end of the period of Roman domination.

157

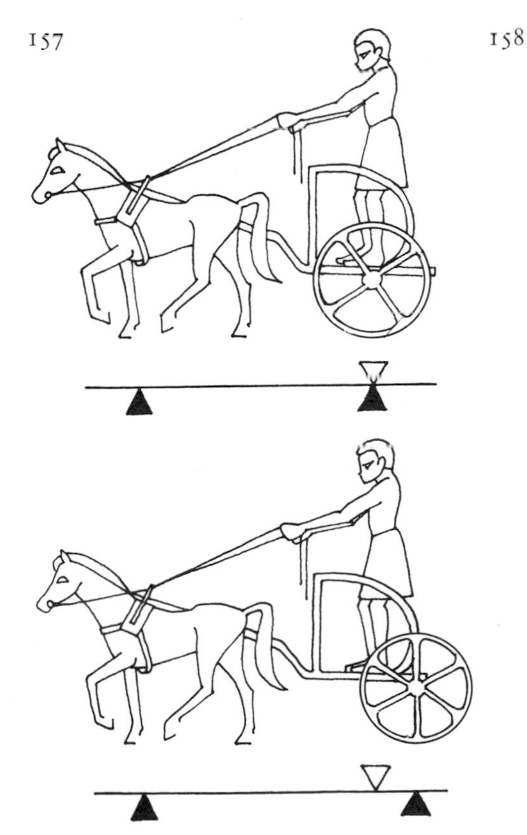

158

With these improvements in harnessing and chariot design one might have expected the chariot to have become a rapid means of communication between one city and the next, but in fact neither in Mesopotamia nor in Egypt were roads taken at all seriously. Although within the cities the roads were often carefully levelled and paved, it would seem that outside the city walls the only roads to which any attention was given at all were those that led either to religious monuments or to quarries or other workings from which heavy burdens had to be carried. Good roadways had not even been developed for military purposes. In records of the period we often read of chariots becoming bogged down and the fighting men having to take to foot and of baggage waggons stuck in the mud. The same attitude equally prevailed when it came to building bridges. While a few bridges were built in or near cities, elsewhere more commonly rivers had to be crossed at fords or by ferries, while for military purposes rafts were frequently used and on more rare occasions pontoon bridges were built across the river.

Thus, by the year 1000 B.C. throughout the larger part of the Near East the seas and the waterways still served as the principal means by which goods were transported over long distances. The wagon remained a slow, lumbering vehicle more suitable for shifting farm produce over short distances than as a means of overland transport, while the chariot despite its improvements could still be used effectively only in open country that was reasonably level.

However, from the same quarter as that from which came the chariot there now arrived a new and more rapid means of getting about, for the people of the Asiatic steppe had learnt how to train horses well enough to be able to ride them. It was in the face of bands of warrior horsemen that the people living round the coastal belt of the Eastern Mediterranean fled southward, only to be repulsed by the Egyptians. Other Near Eastern kingdoms were to suffer in the same way as the peoples of the sea. The kingdom of the Hittites in Anatolia was completely overrun, while another wave of people moved down into Greece and the islands of the Eastern Mediterranean, and while yet a third, amongst whom were the Persians, overran the whole of the Iranian Plateau. Yet a fourth group occupied a large part of the Levantine coast. Only Mesopotamia, at that time under the more or less unified control of the Assyrian rulers in the north and of Babylon in the south, held out against these intruders. Thus the beginning of the first millennium B.C. in the Near East was very largely a period which to the historian and to the archaeologist appears to be so chaotic that for two or three centuries the term 'dark age' seems to be applicable. To those studying the history of technology, however, the period is one of very considerable interest: new ingredients had been added to the pot from which, once it had simmered down, totally new ideas were to emerge.

6
Greeks and Persians

(1000–300 BC)

In the last chapter we saw how an understanding of the working of iron had developed in the mountainous area to the north of Mesopotamia and how, for reasons that are far from clear to us today, this knowledge had not been transmitted to the people of the Mesopotamian Valley or indeed to any other people of the Eastern Mediterranean. However, with the break-up of the old established kingdoms and the scattering of so many peoples the knowledge of iron-working was now carried far and wide. Even so the historian is faced with one of those problems of cause and effect which are so often impossible to answer. The people of the Near East may have taken to the using of iron for the manufacture of tools and weapons because they liked the material and because they found it cheaper to produce than bronze. On the other hand, with the disruption of trade resulting from the destruction of the older kingdoms, iron may have been used in the first place by many people simply as an *ersatz* material in lieu of bronze. Whatever the reason, however, the introduction of iron as a tool-making material had the most enormous effect upon the artisan. Up to this time, for example, the stonemason had had to work with wooden and stone tools for the larger part, but by 500 B.C. there had been evolved a set of mason's tools, largely punches and chisels, which would not seem at all out of place in the hands of a sculptor today. To a lesser degree the same is true of the woodworker. For chisels and saws are better made of iron, to which an edge is more easily put, than of bronze. Indeed, it would be hard to find any craftsman working in the Near East who, by the year 500 B.C., had not to one degree or another taken to the use of iron tools, and in many cases the metal was to completely revolutionize working methods.

During the first three or four centuries following the year 1000 B.C. the old traditions of technology were continued and enlarged largely by Assyria, Babylon and Egypt, and one might truthfully say that during this time the newcomers into the Near East gradually assimilated these technologies from their neighbours. Both the Assyrians and the Babylonians, however, spent much of their time fighting, either endeavouring to enlarge their empires or more simply defending what they already had, and in the field of military technology there is little doubt that the Assyrians took the lead. In a number of early encounters the Assyrians learnt a sharp and painful lesson from marauding horsemen and they very rapidly established their own cavalry. At the same time they appear to have concentrated on building heavier chariots than had been used hitherto: but this may be illusory, for although their chariot-wheels had a far greater diameter than those of earlier vehicles, nearly all the illustrations of Assyrian chariots would suggest that the larger wheels were made by adding a deeper felloe to wheels of the original dimensions. This change presumably allowed them to move over rather rougher ground, and we should also notice that at times the crew of each chariot was as many as four men.

The principal Assyrian military developments, however, appear to have been in the field of siege-craft and in machinery for breaking down city walls. Of these the battering-ram first made its appearance as a wagon covered with shields and with a fixed projection like a pig's snout. The whole vehicle evidently had to be rocked backwards and forwards against the city wall. Such a machine can hardly have been efficient even against walls of mud-brick, and very soon the Assyrians learnt to build an armoured wagon in which the ram, a long beam, was raised and allowed to fall against the wall. The only defence against this vehicle of destruction appears to have been a kind of grappling-iron which was lowered by the defenders of the city in an attempt to immobilize the ram. We have already seen how the Egyptians, some two thousand years before this period, had developed a siege ladder on wheels. The Assyrians, now, probably following the success of their battering-rams, began producing siege towers on wheels, and often incorporated both ram and tower in the same vehicle.

The development of this rather crude military machinery was the direct outcome of the enormous success of the Assyrian army, comprising three main groups; infantry, chariotry and cavalry. Within each of these classes was to be found a further division in which the men were armed either with bows or with spears and swords, so that in any encounter, were it static or mobile, a screen of arrows could be put up allowing those armed with hand weapons to engage an already embarrassed enemy. Against such an onslaught few of the neighbouring states could produce an army able to withstand an

159 Diagram showing the very different stages followed in antiquity to manufacture a spear-head of bronze and one of iron

To a degree the use of iron changed design methods. Hitherto moulds for casting bronze, or the patterns from which the moulds were made, need not have been, and were sometimes clearly not made by the foundrymen. By contrast iron objects had to be forged stage by stage, from a bar of metal, and their ultimate form thus fell within the province of the blacksmith. This distinction, and the effect it had on design, can be seen by examining the stages required in the manufacture of a bronze and an iron spear-head. In the latter case shaping was achieved largely by thinning down and folding the metal when red-hot.

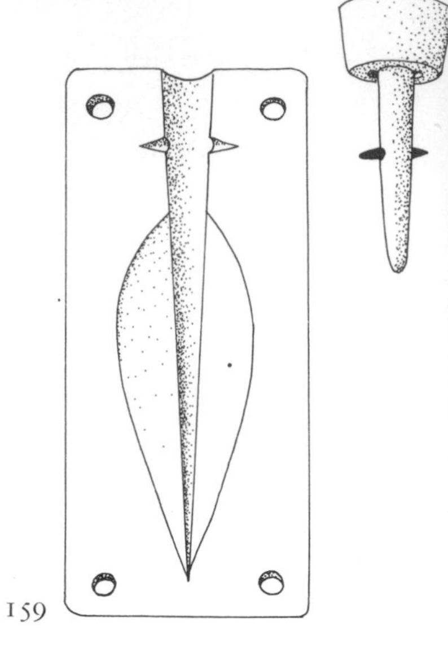

159

160

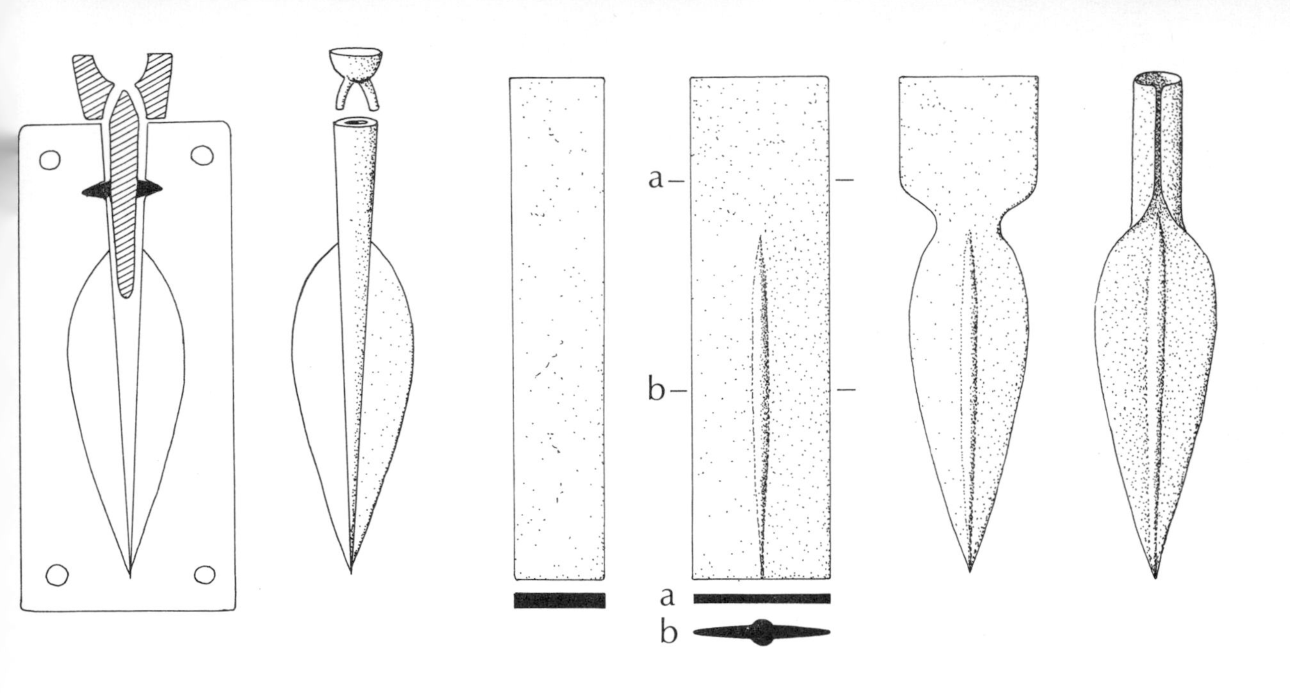

161

160 A Greek bowl of about 400 B.C. decorated in low relief

161 Drawing of a part of the scene from this bowl, showing a blacksmith forging Hercules' club

The general use of iron for making tools and weapons was accompanied by considerable changes in metal-working techniques. The fact that the metal had to be worked when red-hot led to the development of forges with built-in bellows fitted with valves. Presumably the man standing by the forge depicted in this scene is working the bellows, the handle of which is in his left hand. To handle and to shape the metal, tongs and hefty hammers were needed as, too, was a substantial anvil.

162 Diagram showing various methods devised for lifting blocks of stone: drawn from Greek sources, after 800 B.C.

The introduction of iron tools made an enormous impact on many trades and industries, but especially on the working of stone. Iron wedges and hammers made the process of quarrying far simpler, while chisels and punches allowed a greater ease of shaping. Furthermore, because it was now possible to cut exact mortices in stone, it became possible to use a number of different devices by which to lift pieces of masonry, which in turn simplified the problems of building.

163*a* Part of an Assyrian relief of about 700 B.C., showing a heavy chariot

163*b* Reconstruction of this type of chariot

The Assyrian response to the threat of attack by cavalry was to build larger and more rugged chariots. Wheels were given more spokes and enlarged by, apparently, the addition of a deep tyre of wood. The crew was raised to four – driver, shield-bearer, and two archers or men with hand-arms – and the number of draught horses was raised to four. Since the chariots could still be outmanoeuvred by cavalry, they were often given the support of a number of mounted men.

162

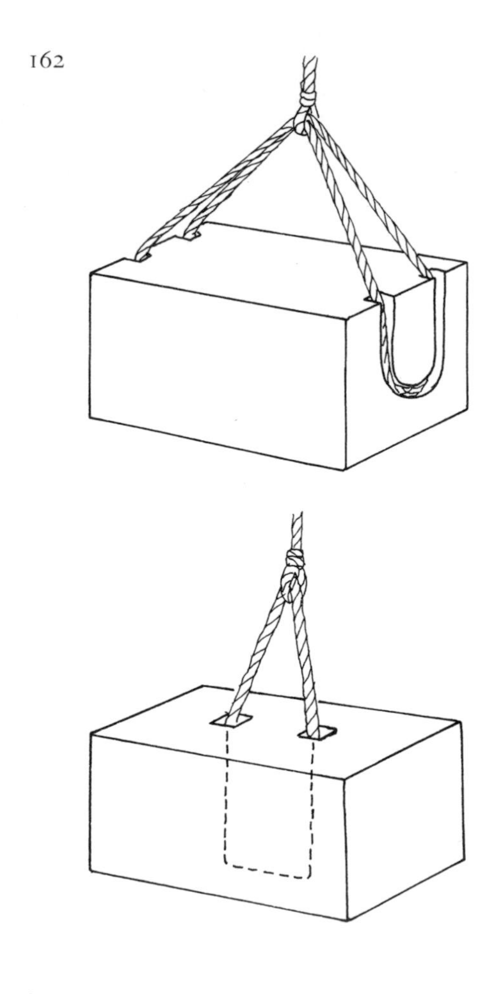

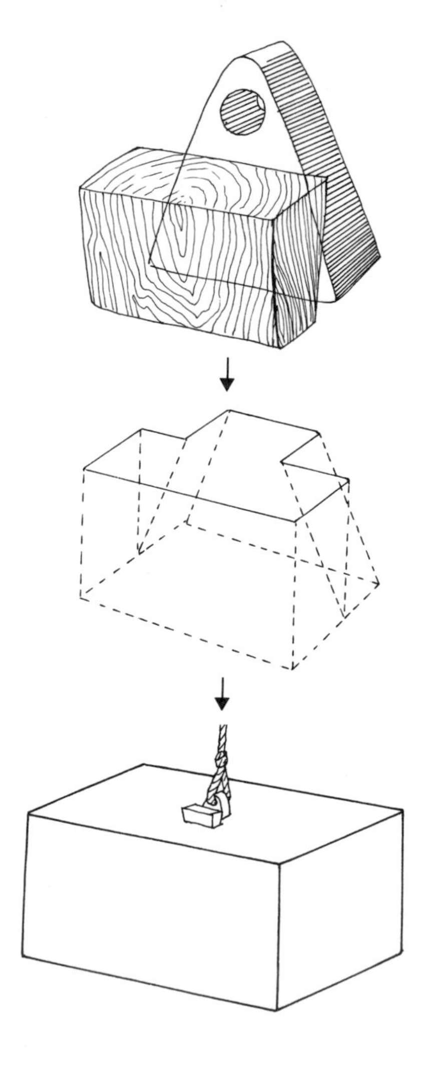

163a

163b

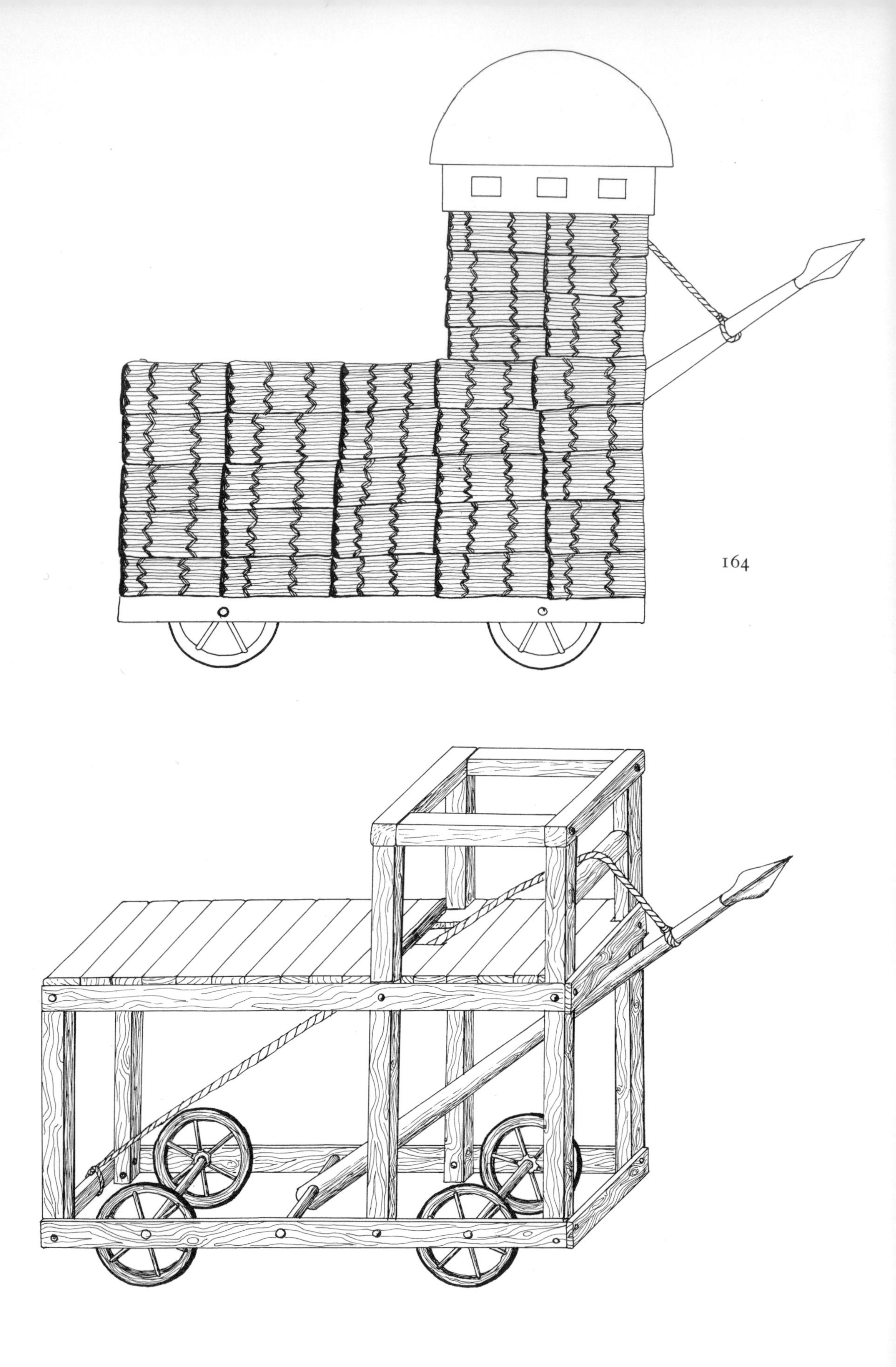
164

164 Drawing from an Assyrian relief of about 800 B.C., and a diagram showing how the siege-tower may have been operated

Unable to withstand Assyria's army, her enemies often took refuge behind city walls. The Assyrians, however, designed a number of siege-engines which were not infrequently depicted on their reliefs. In its most sophisticated form the siege-engine appears to be a combination of a siege-tower and a battering-ram mounted on wheels. The battering-ram appears to have been in the form of a pointed arm that could be used to prise out the mud-brick from the city wall, while archers on the platform above protected those below demolishing the wall.

165 Part of an Assyrian bronze relief of about 900 B.C., showing a pontoon bridge

166 Part of an Assyrian relief of about 800 B.C., showing troops crossing a river on inflated skins

165

166

167

168

167 A gufa – the biblical coracle – in use on the Tigris today

168 Inflating a goat-skin float, Nepal

For crossing rivers the Assyrian army resorted to a number of strategems. At times inflated skins were used as floats, and equipment, such as a chariot, was taken to pieces and carried across the river in sections. Elsewhere a pontoon bridge would be constructed using as floats large basket-work boats made watertight with bitumen. Curiously, both inflated skins and the simple basket-work boats – gufas – are still in use in the Near East today.

Assyrian attack in open battle, and their opponents had to take refuge either in cities or in the mountains. It was indeed in the eastern mountains of Persia that the Assyrians met with perpetual and incessant reverses, and it was from this quarter that military disaster was ultimately to overtake them.

It would be unfair to suggest that the Assyrians' only interest in technology was in the military field, for having once conquered a people, they had developed the habit of removing their more useful citizens, particularly their craftsmen, to other centres, with the result that Assyrian cities literally teemed with artisans drawn from other parts of their empire. This mixing of craftsmen from different regions should have resulted in cross-fertilization and in technological improvements; but, while they excelled in the making of glassware or glazed bricks, no really new technologies seemed to have developed under the Assyrian regime, and it is to the east and to the west of that country that we must now look.

By about 800 B.C. the situation in Greece and the Aegean islands had settled down. The newcomers, the Hellenes, initially an agricultural people, had learned much from the original population whom they had clearly not totally eliminated. From our point of view the most important single technique that they mastered was how to build and to sail ships. Their vessels appear to have differed very little from those of the people they had conquered. Although they chose to depict them in a rather different manner, the Greeks are to be seen either rowing the identical low craft with a ram bow and a high upturned stern, or sailing the same type of craft with high stem and stern as had their predecessors in this area. These vessels were to result in the Greeks not only becoming merchantmen but also, later, as their population expanded, great colonizers. This in turn brought them into contact with all the peoples not only of the Eastern but also of the Western Mediterranean.

169 Reconstruction of the Greek 'long-boat' of about 600 B.C. made from a drawing on a vase in the Louvre

This drawing of a Greek 'long-boat' is probably about as near as one can hope to approach to a blueprint. The picture is part of a scene on the 'François' vase in the Louvre and is unfortunately not complete, a small section being missing just abaft the bows. When reconstructed the vessel appears to have been between thirty-five and forty feet long with a freeboard of about eighteen inches, a ram bow, a high stern, and an awning in the forepeak. The hurdle-work above the gunwale suggests a crew of twenty-six oarsmen. The original drawing makes it perfectly clear that the oars were placed immediately over the gunwale. If, on the other hand, a spare crew were seated inboard of the oarsmen, each man provided with an additional longer oar, both crews could row simultaneously in times of stress, the inboard crew rowing over the hurdle-work. This would have brought the total crew up to fifty. In fact this vessel is probably the penteconter, the fifty-oared long-boat, known otherwise only from literary sources.

The original drawing was made shortly after 600 B.C., but earlier less reliable drawings suggest that for at least two centuries before, vessels of this general character had been in use amongst the Greeks.

The shipping trade in the Mediterranean, however, was not the sole monopoly of the Greeks, for they shared it with the people of the Levant, particularly the Phoenicians, who themselves were equally great colonizers, Carthage on the North African coast being the most successful and best known of their settlements. But the Phoenicians differed from the Greeks in one important respect – they were Asiatics and their trade was carried out with and for Asia, with Assyria and other people in the eastern end of the Mediterranean. To the Phoenicians, presumably, Assyria and its neighbouring countries were the centre of the world, and for them, except in matters of shipping, it was unthinkable that anybody elsewhere could produce manufactured wares or goods better than they. By contrast the Greeks were evidently not quite so convinced of Assyrian technological supremacy, and by the end of the seventh century B.C. the Greeks were already demonstrating their potential superiority in many fields.

The most obvious example of the advances made by the Greeks in the technological field is that of their pottery. The superiority of Corinthian and Attic wares to anything being produced elsewhere in the Western World at this period is without question. But this pottery did not just happen. Behind it lay a number of technological innovations that the Greeks clearly devised themselves. To begin with the potter's wheel, this was no longer a small turn-table placed close to the ground but had become a large fly-wheel with a working-head raised above it to a height of a foot or eighteen inches above the ground. Sitting on a stool, the potter shaped his vessels on the wheel-head while an assistant, probably an apprentice, sat opposite turning the heavy fly-wheel by hand. This heavier and therefore more steady wheel, we shall see, was to be of considerable importance later. Secondly, once the pottery had been shaped and dried it was replaced on the wheel, probably inverted, and shaved, thus producing a far finer surface than would have been possible simply by throwing on the wheel-head. This habit of shaving pottery does not appear to have been carried out elsewhere at this time, and this practice was probably to have an effect in a completely different field of technology. Finally, the contrasting colours of black and red were produced by a very sophisticated process

170 Limestone figure of a potter from Cyprus, about 600 B.C.

171 Reconstruction of the Greek potter's wheel of about 600 B.C., made from a number of painted vessels of this period

Even as late as the sixth century B.C. Greek potters depicted themselves seated on a low chair, shaping their wares on a large turn-table with a raised central disc on which the clay was worked. Either the potters turned the wheels by hand themselves, or a youth sat on a stool opposite to perform this duty. So far the wheel showed very little change from that used far earlier in Egypt. Unfortunately, shortly after this date potters no longer depicted themselves on their wares, with the result that we do not know precisely when the kick-wheel was introduced.

170

171

172

173

demanding the production of an extremely fine clay material followed by a fairly elaborate sequence of firing. The clay slip used for producing the black areas of this pottery was made by stirring clay with water and an alkali, probably leach from wood ash, and allowing the material to stand, when the very fine clay particles alone would remain suspended. This suspension was then decanted and allowed to evaporate until it achieved the required consistency with which to paint on the decoration. The wares were fired in a domed kiln to a temperature little short of 1,000 °C and at this point the openings of the kiln were closed down, the momentary effect being to blacken the entire surface of the wares. When the kiln had cooled to a little below 800 °C the apertures were reopened, so readmitting air, with the result that those areas which had been covered with the fine slip remained black while those left untreated gradually lost their blackness and became a clear red. Stated in this bald way the process seems simple, but to achieve the end result a very great deal of observation, trial and error were necessary, and indeed it is only in recent years that the method used has been re-established by scientists anxious to determine how it was done.

172 Reconstruction of a Greek kiln of about 500 B.C., made from a number of decorated plaques from Corinth

173 Fragments of a decorated plaque from Corinth with the remainder of the scene reconstructed

At first sight Greek potters' kilns seem to show little change from those in use in Mesopotamia two thousand years before (Figure 51). On the other hand, the fire opening had been extended, presumably to allow a larger space for combustion and to improve draught. A loading door was usually shown with a spy hole through which the potter could see what was going on inside the kiln. Frequently the potter depicted himself climbing on to the dome of the kiln to open or to close the top vent, or raking out the fire, vital steps in the production of black-and-red wares.

Corinthian and Attic wares were traded widely throughout the Mediterranean and ultimately found their way even into Central Europe. Although we have no idea who developed this interesting system of pottery-making and decoration we are more fortunate in other fields, for we come now to a period in which historical records begin to be more reliable. In the small Greek states on the western coast of Anatolia, particularly in Ionia, is recorded a group of people who appear to have devoted themselves whole-heartedly to technological improvements. Of these Thales of Miletus is probably, undeservedly, the best known. In his travels Thales had evidently seen Egyptian surveyors at work and had studied their methods of land measurement and the equipment they used. It was also probably from this source that he learned of astronomy and how the Egyptians used the stars to determine positions. Armed with this knowledge and a great deal of practical common sense, Thales devoted himself to the study of navigation at sea. By a system of triangulation, in which, of course, the magnetic compass was not used, he evolved a method of determining the distance of ships at sea from dry land, and from his knowledge of astronomy he laid at least the foundations of navigation by the stars.

A contemporary of Thales, Anaximander, was evidently equally interested in navigation for he allegedly produced the first map of the world. Maps of one sort or another had existed long before Anaximander's day. These were often crude estate maps or merely sketches to show how to get to such and such a point, and although they often stated distances and areas they were seldom drawn to scale. The known world in Anaximander's day was admittedly not very large, but nevertheless this was for one man a gigantic undertaking and as a concept an enormous step forward. Like Thales, Anaximander was also interested

174 Reconstruction of a gromar from fragments found at Pompeii

Simple surveying equipment such as this, which allowed the setting-out of accurate right angles, had long been in use in Egypt. It was, however, the Ionian Greeks who first appear to have used surveying equipment for the purpose of navigation.

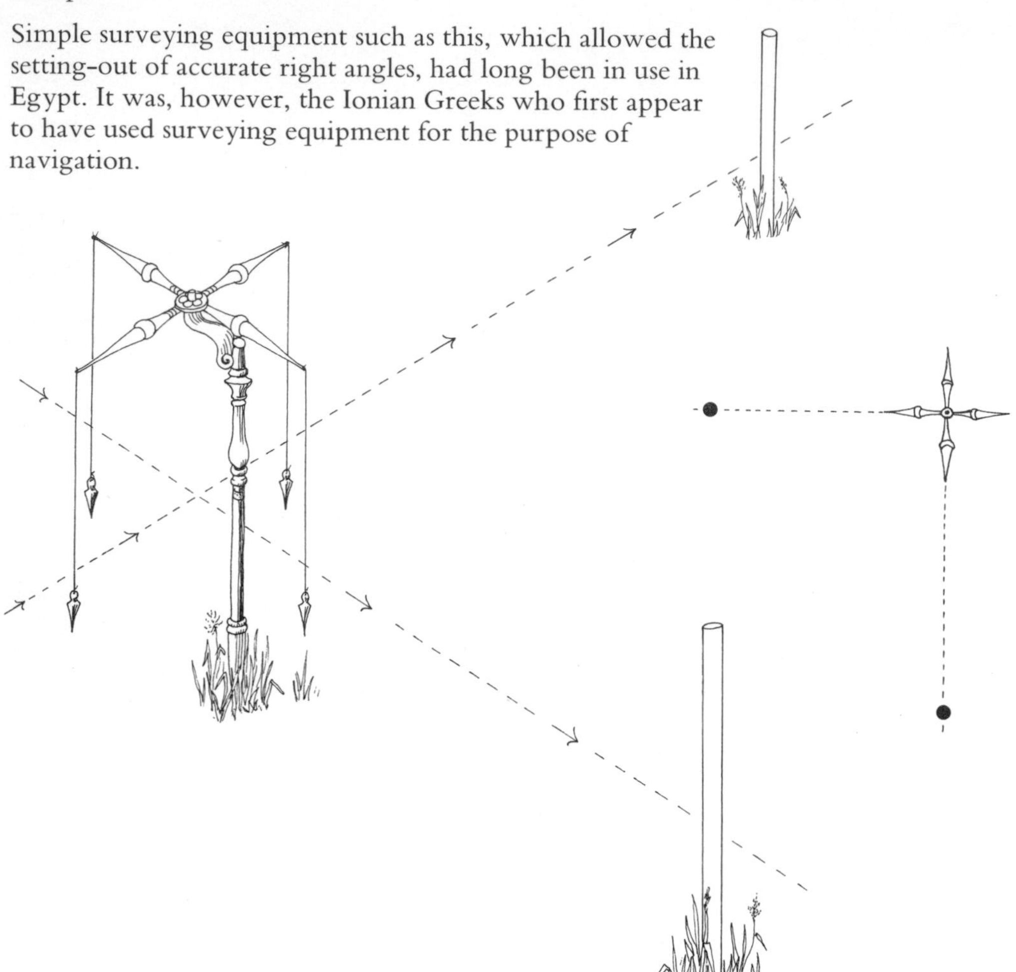

in the behaviour of the heavenly bodies and it is to him, too, that is given the credit of inventing the first 'sundial', more probably a simple astrolabe. Once again, Anaximander was clearly drawing on knowledge that already existed, for sundials of a fairly simple character were in use elsewhere before his day, but he attempted to give the device some sort of correction for the varying seasons, so providing a useable time-piece. Two other inventors of this period – Anacharsis the Scythian and Theodorus of Samos – have been less kindly treated by historians than Thales. Between them they were credited with improving the anchor, the bellows and the potter's wheel; of inventing the lathe and the key; and of improving a number of mathematical instruments as well as devising a method of casting bronze. This is an impressive list that would have done credit to, and made the reputation of, any inventor. The difficulty in accepting the statement lies in the fact that many of the objects and methods in this list were known to exist before the lives of Anacharsis and Theodorus, and it has been assumed these men merely filched ideas from Egypt, Assyria and elsewhere and brought them back to Greece. To some extent this may have been true, but nevertheless the evidence suggests that the improvements listed were made during this period. We have just seen that the Corinthian potter worked on a wheel which was fairly close to the ground, requiring an assistant to make it turn continuously. Following this we have no other evidence of what potters' wheels looked like until the Ptolemaic period in Egypt some three hundred years later, when we find the potter seated on a high bench, the wheel-head raised to waist height and the potter himself rotating the wheel by kicking it with the right foot.

175 An Egyptian relief of about 300 B.C. showing the god Khum seated at a kick-wheel

176 Reconstruction of this type of kick-wheel

177 Pottery figure made by John Broad in 1883

By 300 B.C. the working head of the potter's wheel had been raised as, too, had the height of the potter's seat. He could now sit at the wheel and kick it with one foot to make it rotate, so leaving both hands free to shape his wares. These changes in design were possibly the work of Anarcharis, who is alleged to have 'invented' the potter's wheel. With very little modification this type of wheel was still in use in Europe as recently as the end of the last century.

175

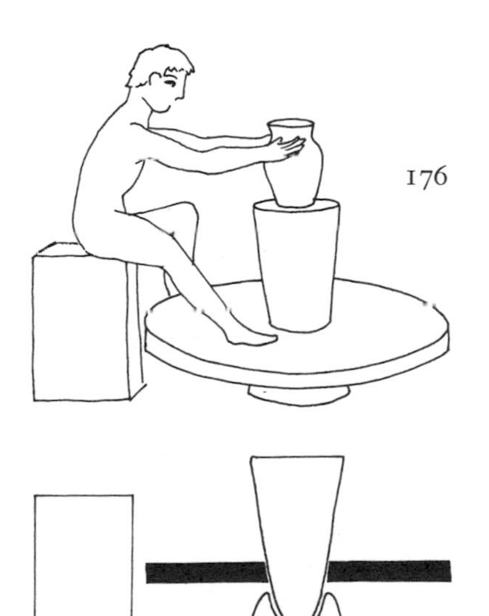

176

177

178 A lathe as depicted in an Egyptian low relief of about 300 B.C.

179 Reconstruction of this type of lathe

180 A similar simple lathe as used by the Hunza today

181 Diagram showing how the lathe was adapted for stone and glass cutting by the addition of a lap wheel

The lathe, the invention of which was again attributed to Anarcharis, is first depicted in the third century B.C. in Egypt. The object being turned and the spindle were held horizontally between stocks. A cord was passed round the spindle and one operator made the spindle and workpiece rotate back and forth by pulling on the cord. A second artisan cut back the workpiece with a chisel. The idea of the lathe may well have been derived from potters shaving down their dried wares on the wheel.

It seems, therefore, quite possible that Anacharsis saw the Corinthian potter's wheel, realized the illogicality of the device and suggested certain improvements. By raising the wheel-head and the height of the seat the potter was able to make the wheel rotate by kicking it himself. Anacharsis is also said to have improved the anchor. Certainly up to this point the only anchors known were large stones drilled at one end to take the mooring-rope and at the other end to take short wooden flukes. Such anchors would have dragged in heavy seas. After this period we find in use the conventional anchor with developed flukes and, once again, it seems probable that Anacharsis saw that an anchor could be prevented from dragging by redesigning it in the form of a grapple, with which he may well have been familiar. His third invention, the bellows, already existed as hand- or foot-operated skin bags, or as skin-covered drums. After this period, however, we find that the bellows with a clapper valve was in use, and once again there is no reason to believe that Anacharsis was not responsible for the innovation.

The earliest lathe that we know of comes, again, from a Ptolemaic tomb painting, and thus when we read of Theodorus of Samos having invented the lathe there seems no good reason to doubt this statement, especially when we recall that he may have seen Corinthian and Attic potters shaving their wares on the wheel-head, a process which could easily have suggested to him how a piece of wood might be shaved down in the same way were it made to rotate. It is more difficult to say what is meant by the statement that Theodorus invented a method of casting bronze. But yet another Ionian inventor, Glaukos of Chios, is reputed to have invented the welding of iron, that is to say, the joining

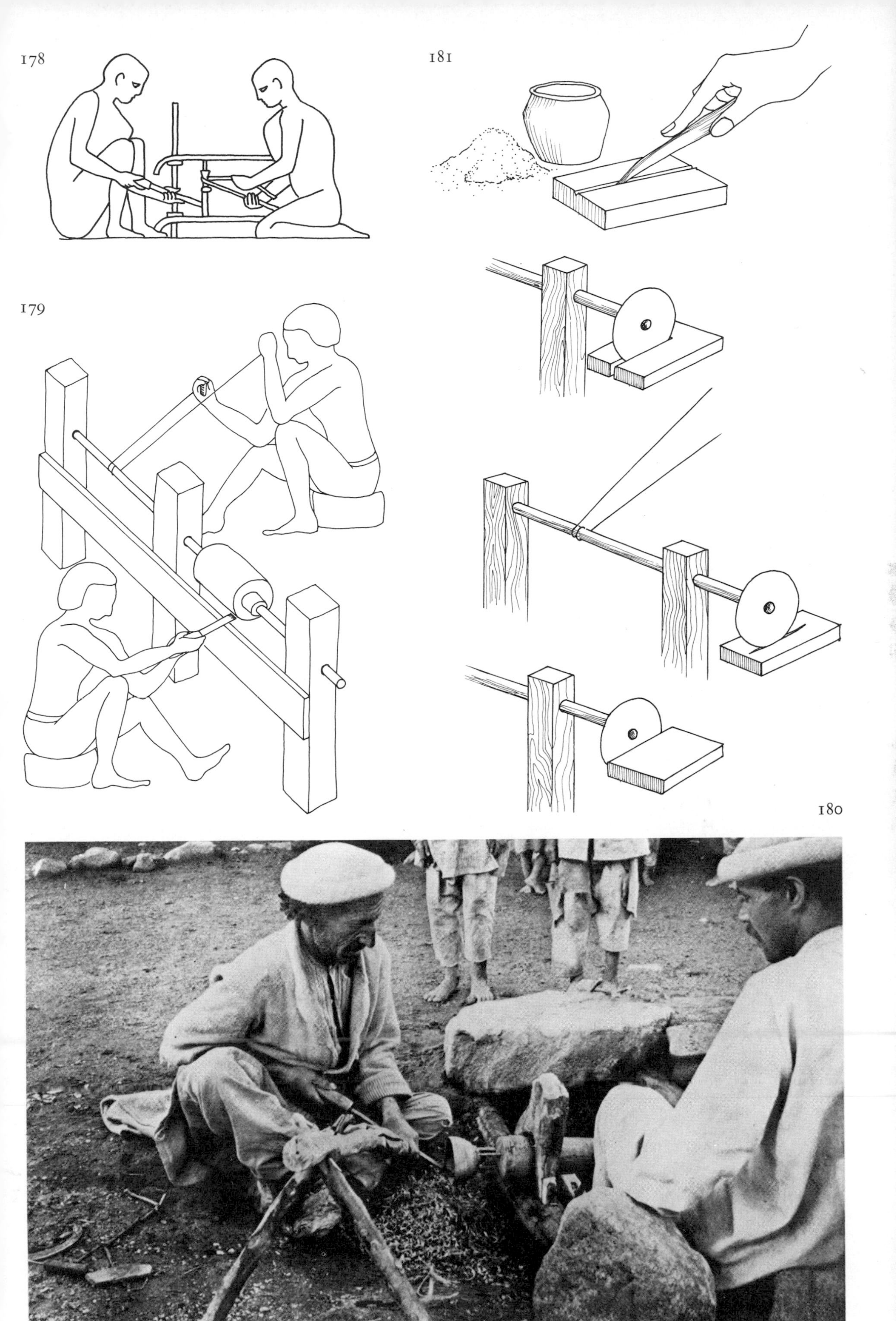
178
181
179
180

182 A Greek vase of about 600 B.C., showing an anchor being thrown overboard

183 Another form of stone anchor in use from about 1000 B.C. to which wooden flukes were probably fitted

184 Anchors as depicted on lead weights from Syria, about 200 B.C.

In the seventh century B.C. many ships' anchors were still little more than large stones drilled through the middle to take the hawser. Anchors of this kind must have dragged in heavy seas. To overcome this defect an anchor in the form of a triangular stone was developed. The hawser was tied through a hole in the upper angle of the stone, while a pair of wooden flukes were fitted through holes near the lower edge. Traditionally the anchor, as we understand it, was invented by Anacharsis, and certainly by the second century B.C. the anchor had attained the form in which we know it today.

of one piece of iron to another by heating to a good red heat and hammering on an anvil. Certainly until this time blacksmiths had avoided this method and had often gone to peculiar extremes in order to join iron by means of such devices as rivets and lapped flanges. But with the development of the bellows, allowing a good red heat to be achieved easily, coupled with the development of really heavy anvil tools (as depicted on Attic and Corinthian vases) there is no reason to doubt that Glaukos may well have improved the technique of welding. Indeed, the evidence suggests that the Ionian Greeks were giving all the processes of metal-working a critical eye, the results of which are very evident. Their bronze statuary, for example, apart from any aesthetic qualities, shows a considerable technological advance over anything being produced elsewhere in the Near East at this time, being made of fewer and larger castings than hitherto, and incorporating such devices as iron armatures for its support.

What is, however, more interesting about the Greeks of this period is not whether or not certain inventors were responsible for what has been claimed for them, but the mere fact that their names are recorded as having made those inventions. Plainly the attitude of the Greeks to technological innovation as reflected in these statements is very different from that of the Asiatics, amongst whom a craftsman remained for all his skill and all his inventiveness a mere craftsman. That this attitude was short-lived we shall soon see, but for a few centuries at least the status of the technician amongst the Greeks was raised far above that of his fellow technicians in the Asiatic countries. To be interested in technology was respectable and an inventor was regarded as a benefactor. The Greeks, however, did more than metaphorically raise the level of technology, for they raised it physically as well.

A study of pictures of craftsmen in early Egypt and Mesopotamia will show them either standing at their work or kneeling or squatting on the ground: essentially the ground was their work-bench. By contrast, when the Greek artisan is not shown standing, more often than not he is to be seen seated on a stool or a chair, his work being carried out on a low bench or table. At first sight this may seem insignificant, and one could argue that environmental differences, the kinder weather in countries other than Greece, would account for the choice of working position. This may well be true, but it does not alter the ultimate fact that many crafts are far better carried out on a bench than at ground level, as for example wood-working, masonry and fine metal-working, to mention but a few. Indeed it seems unlikely that many of the iron tools evolved during this period for particular

185 Egyptian sandal-makers as depicted on the walls of a tomb of about 1500 B.C.

186 A Greek sandal-maker of about 600 B.C., from a vase

187 A shoe-maker in Jordania today

Unlike his Egyptian and Asiatic counterparts the Greek artisan does not seem to have enjoyed working while squatting on the ground. More often than not he is shown seated on a chair and working at a table or bench. As has just been seen in the case of the potter's wheel, this difference in posture allowed the development of tools and techniques that would have been impossible had the working position been one of kneeling or squatting. Throughout a large part of Asia and Africa today craftsmen still show a preference for working close to the ground.

185

186

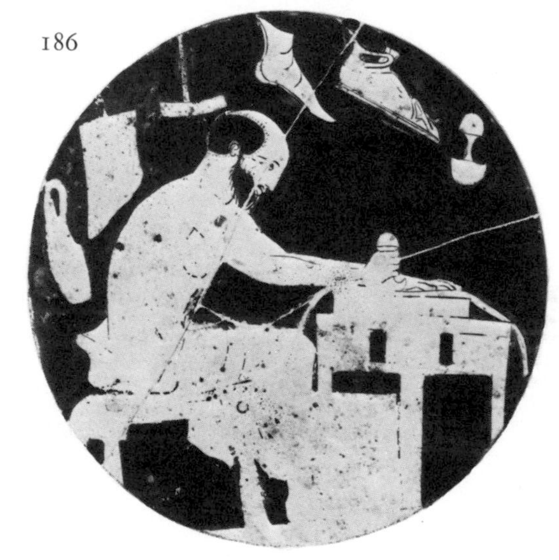

187

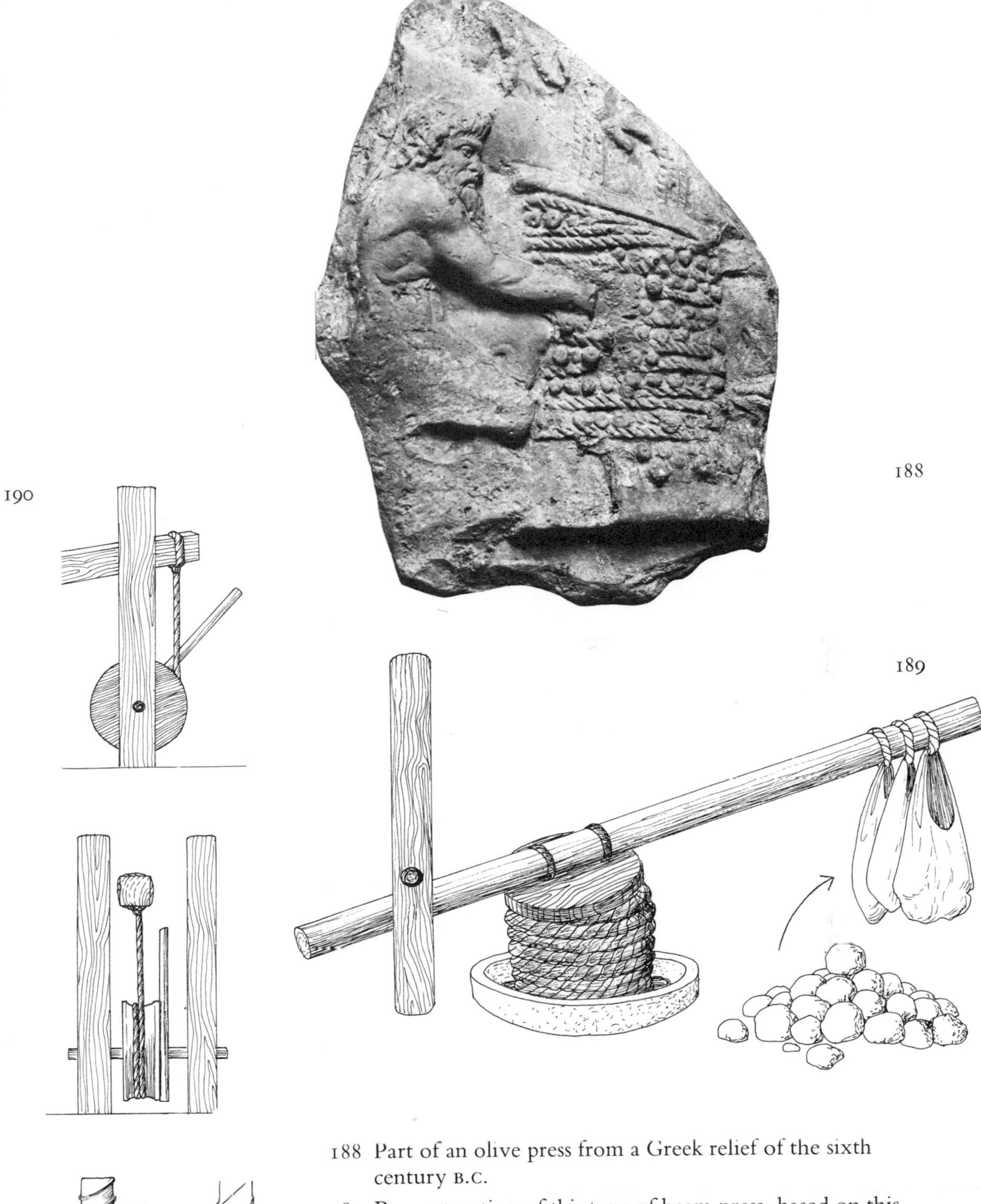

188 Part of an olive press from a Greek relief of the sixth century B.C.

189 Reconstruction of this type of beam press, based on this relief and a number of painted vases

190 Diagram showing how the pulley and the screw were later used to work the beam of this kind of press

Olive oil was to become one of the main Greek exports. Initially the oil was extracted in a simple beam press, of which a few illustrations are known. Thales, who was in reality the father of navigation, is also remembered as once having 'cornered' the market in olive oil.

(a)

 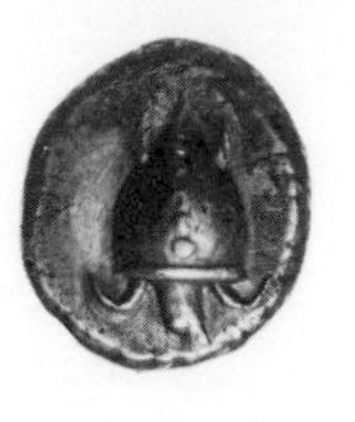

(b)

(c)

191 Coins (staters) from (a) Ionia, about 600 B.C.; (b) Aegina, about 575 B.C.; (c) Thasos, about 525 B.C. (a) is made of electrum; (b) and (c) of silver

With the growth of trade in such commodities as pottery and olive oil, the demand grew for an acceptable token of exchange. Coins, pieces of metal of guaranteed weight and purity, stamped with the mark of the issuing authority, first appeared in the seventh century B.C. Early coins were fairly simple in design, but by the sixth century most countries of the Eastern Mediterranean were minting their own coins, the designs of which showed increasing complexity.

crafts could have been developed at all had the craftsman continued to work at ground level.

During the period immediately preceding the seventh century B.C. the Greeks had, agriculturally speaking, slowly adapted themselves to their new environment and had developed the cultivation of the vine and olive to such a degree that olive oil and wine together with their pottery became their principal exports. There can be little doubt that it was this enormous volume of trade that virtually forced the people of the Mediterranean into adopting a medium of exchange, rather than continuing to base their commerce on direct barter. Reputedly, Croesus of Lydia, a small state in northern Anatolia, introduced a form of currency in about the year 700 B.C. This ruler introduced the first true coins, pieces of metal stamped with a guaranteed weight and purity. A hundred years later the Greek cities were all issuing their own bronze and silver currency, a practice that was soon to spread throughout the whole of the civilized world.

One cannot, of course, issue currency of guaranteed purity unless one has metallurgists capable of producing pure metals. Nor can the coins be of a guaranteed weight unless one has weighing devices capable of weighing accurately. These last two observations may seem rather trite, but they do show, nevertheless, how far the Greeks had advanced by this period. To be able to produce pure silver coinage they must have been familiar with methods of cupellation, in which the impure silver was heated either in a furnace or in a crucible with materials capable of absorbing such impurities as might be present, principally lead. From a slightly later period we have a number of descriptions of this process which show that by then it was old and well established. Furthermore, by cupellation with common salt, silver could be extracted from electrum, the naturally occuring gold-silver alloy, so allowing the production of reasonably pure gold coins.

The development of a monetary system speaks for itself. The volume of trade being carried out by the Greeks and other countries of the Near East was obviously on the increase, and so far as the Greeks and the people of the Levant were concerned, this trade was being conducted by sea. We might reasonably expect, therefore, to find at this period that the Greek shipping and that of the Phoenicians was pretty well identical. But this is far from being the case, and the very differences themselves say a great deal about the attitudes of these two different peoples. By about 700 B.C. the people of the Levant coast had made one very important change in the placing of the oarsmen in those ships that were rowed. Hitherto there had been a single line of rowers, the oars being placed over the gunwales while the rowers were protected by wash-boards, as we have already seen. In time the wash-boards became incorporated in the hull of the ship which was then

192 A stone relief from Southern Turkey of about 800 B.C., showing a warship during battle. It is manned by a single bank of oarsmen

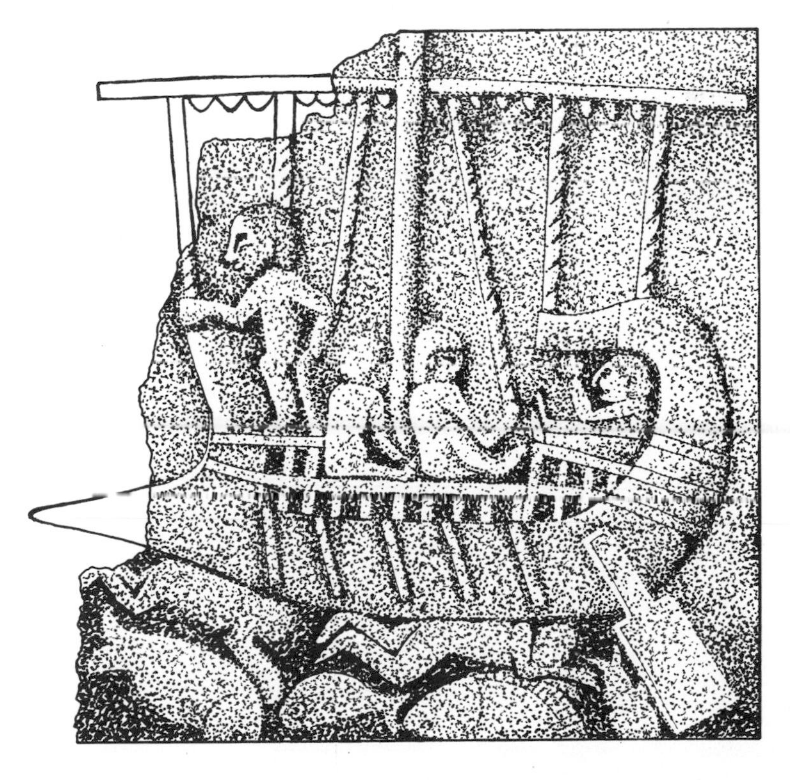

193 An Assyrian relief of about 700 B.C., showing a ship of the same general form, but manned by two banks of oarsmen

194 Conjectural reconstruction of an Assyrian bireme of about 700 B.C., based on this and similar reliefs

During much the same period that the Greeks were developing the penteconter, the people of the Levant made a number of changes in the design of their ships. The wash-boards, in use well before 1000 B.C., were incorporated into the ship's side so that the oarsmen were in effect rowing through a line of ports. A second bank of oarsmen were then able to row over the gunwale. This two-bank arrangement, the bireme, was a permanent one unlike that suggested for the penteconter, which was presumably inspired by the Levant biremes. The hull appears to have been shorter than that of the Greek vessels, and a superstructure was added to it to give advantage to the armed men. This gives the vessels the appearance of having been unduly top-heavy.

built with a row of portholes through which the oars were placed. More simply, the ship's sides were extended upwards and rowing took place through ports. But the disadvantage of this system was that the number of oarsmen was restricted by the length of the vessel. The Phoenicians, however, found that they could increase the number of oarsmen by placing a second row of seamen slightly above and inboard of those rowing through the ports, their oars protruding over the gunwales of the extended ship's side. This system virtually doubled the number of oarsmen without extending the length of the ship, and one rather suspects that this scheme was the outcome of a policy which demanded that the keel should be a single long timber. There is a sufficiently large number of pictures of galleys of this type to suggest that the number of oars employed on each side, even in double banks, never exceeded twenty, and that the vessels were frantically top-heavy, especially when they were used as men-of-war, when a deck had to be provided at a raised height to allow the archers and other armed men to be able to fire upon their enemies. Furthermore, it seems unlikely that such ships with their severely limited number of oarsmen and their great top-weight, were able to operate in heavy weather. By contrast, Greek vessels of the same period seem to have become excessively long and rather low in the water, frequently being depicted

195 A Greek bireme as depicted on a vase of about 500 B.C.

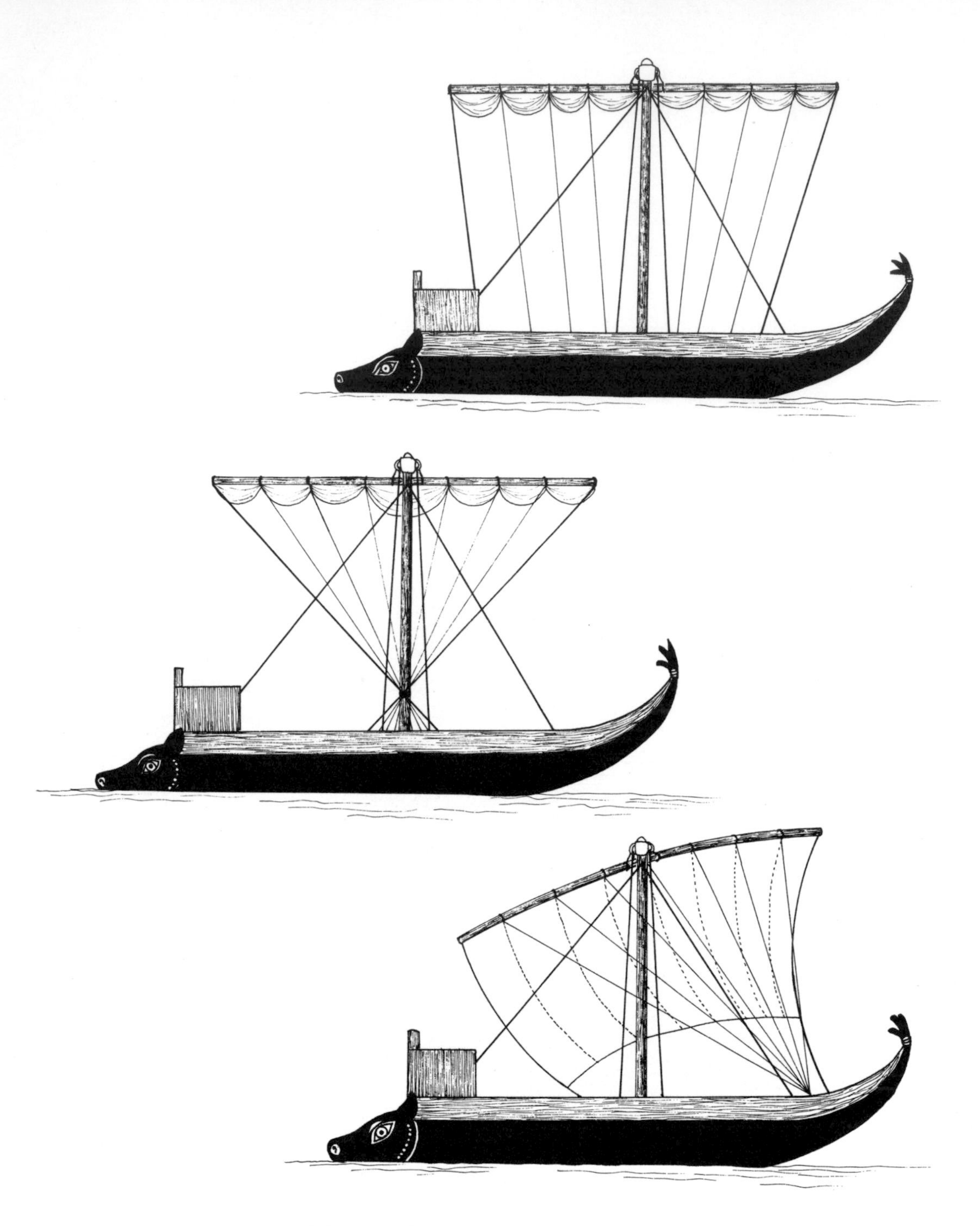

196 Diagram showing the rigging used on Greek and Levantine shipping of this period, based on reliefs and painted vases

By 500 B.C. the Greeks, too, were building biremes with one bank of oarsmen rowing through ports and the second over the gunwale as in the Levantine manner. Illustrations of rigging of this period show that the Greeks and Phoenicians were using identical systems. Unlike earlier Egyptian sails, furling was now done by hauling on sheet-lines which were secured either to the foot of the mast or to the gunwale. The yard-arm could be lowered by means of a pair of hoists that ran through a block at the mast-head.

with twelve or thirteen men rowing on each side, the oars projecting over the low gunwales. Above the gunwale is often shown a light framework of wood and, in early paintings particularly, this is distinctly shown as being lashed together. This framework would be quite useless as a protection against the weather and, in any case, is never shown covered with an awning, although other parts of the ship where awnings are used are quite clearly illustrated. We know from literary sources that it was at this time that the Greeks introduced a vessel that they called the penteconter, the fifty-oared vessel. It seems likely, therefore, that the ships that we see depicted with twelve or thirteen oars on each side were normally rowed by only half the crew, while the other half were resting. In times of stress, however, when great speed was required for fairly short periods, the resting crew, seated as in the double-banked Phoenician ships, could put out rather longer oars over the hurdle work, doubling the number brought into play, and thus raising the total number of rowers to fifty. Unlike the Phoenician vessels, such a ship would have been capable of countering the strong currents of the Hellespont and so entering the Black Sea, and it is important to notice that it is from this period onwards that Greek trade with the Black Sea coasts began to take on a major role in the country's economy.

At this time the Phoenician ships were rowed by slaves, and their cramped conditions mattered little to their masters. By contrast, Greek shipping was manned by free men and the conditions under which they worked were probably the subject of debate, a matter at which the Greeks then, as now, excelled. Significantly, when the first pictures of double-banked ships of Greek design first appear, around 500 B.C., slavery had already become a major aspect of Greek economy, although the oarsmen were still, technically, freemen.

We must now turn our attention to the progress of affairs on the mainland of the Near East. In 625 B.C. the Babylonians defeated their Assyrian neighbours to the north. It may seem strange that the Assyrians with their magnificent war machine were defeated by their apparently more puny neighbours, but in fact the Babylonians cannot take all the credit for the defeat for they were helped in no small degree by newcomers into the district, an alliance of tribesmen, lightly armed, but superb horsemen, who formed a core of light cavalry with which the Assyrians found themselves unable to cope. Amongst these people were the Medes who were to figure later on in the history of this part of the world. But for the moment this change in rulers made very little difference. Babylon continued the traditions of Assyria, and amongst the people moved into captivity, many of whom were craftsmen, were the Jews. The Babylonian Empire, however, was not to last long, little over a century, in fact, for in 538 B.C. the Babylonian rulers in turn were overthrown. A second group of tribesmen, equally expert horsemen, the Persians, in alliance with the Medes, overran the whole of Mesopotamia having first dominated the Iranian Plateau. Hence, the whole of Mesopotamia, Persia and a large part of what is today Turkey fell into the hands of a single group of rulers.

It is very easy to underestimate the contribution the Persians made to technological advance. They had become in their outlook typically Asiatic and politically they ran their empire in much the same way as had the Assyrians and the Babylonians before them. It was, however, a vast area stretching ultimately from the Punjab to the Mediterranean, and possibly because of this the Persians made one very great contribution to mankind's welfare. They developed a system of communications based on roads maintained by the central authority. These roads had posting stations at regular intervals and by virtue of this messages carried by relays of mounted horsemen could be transmitted over great distances in remarkably short spaces of time. Many of these roads were paved, although, of course, in mountainous areas there were long stretches that were not. However, they all shared one thing in common; they had to be maintained in a state that would allow horsemen to move along them rapidly. While by today's standards we might not think very highly of these roadways, they were, nevertheless, a vast improvement upon the haphazard trackways and by-ways with which previous rulers of these areas had been content.

By 500 B.C. the stage was already set for a major conflict. On the one hand were the Persians ruling a vast Asiatic empire centrally controlled and capable of mustering enormous technical resources. On the other hand were the Greeks, a young, vigorous, sea-going people, technologically on the move, anxious to increase their trade and, because of their rising population, creating more and more colonies. It was only natural that the Greeks who had already settled on the western coast of Anatolia and the islands close to it should revolt against a Persian overlordship. Much of the first quarter of the fifth century was, therefore, devoted to resolving the rivalry of these two great peoples. It ended essentially in a stalemate. The Greeks beat off a threat of invasion from Persia. The Persians found themselves unable to curb Greek commercial enterprise.

If this war can be said to have achieved anything at all, it acted as a stimulus to the Greeks to build better shipping. As early as 600 B.C.

197

198

197 Drawing made in the nineteenth century of a relief from the Acropolis (since destroyed), showing a section of a trireme of about 500 B.C.

198 Modern model reconstruction of a Greek trireme based on reliefs and literary sources

199 Clay model of a trireme from Egypt, probably about 500 B.C.

200 Reconstruction of this type of trireme

Triremes, ships with three banks of oars, were being built in Egypt and the Levant by 600 B.C. A century later the Greeks were building them, too. We know very little about their construction. Literary references give the overall dimensions and the number of oarsmen while marble reliefs, such as that from the Acropolis in Athens, show the spacing of the oarsmen. Students of early shipping have suggested that the upper bank of oarsmen seen in the drawing of the Acropolis relief were rowing through an outrigger, and the model in the Science Museum has been made to incorporate this feature.

The clay model from Egypt suggests that in countries other than Greece triremes were not fitted with outriggers. The reconstruction shown here assumes that the Levantine trireme was little more than a bireme (see Figure 193) with the addition of a third bank of oars.

199

200

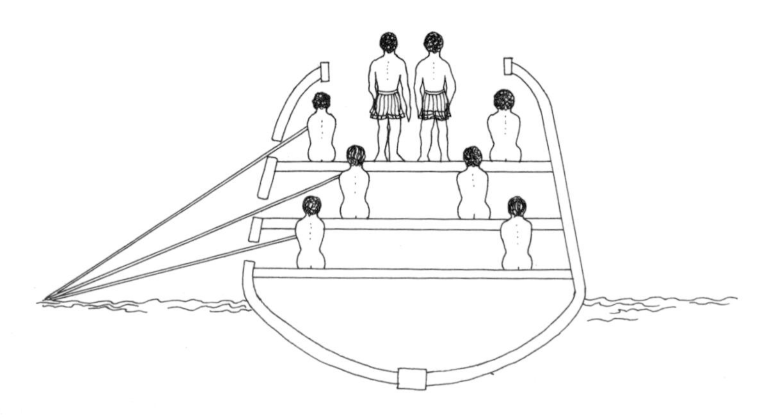

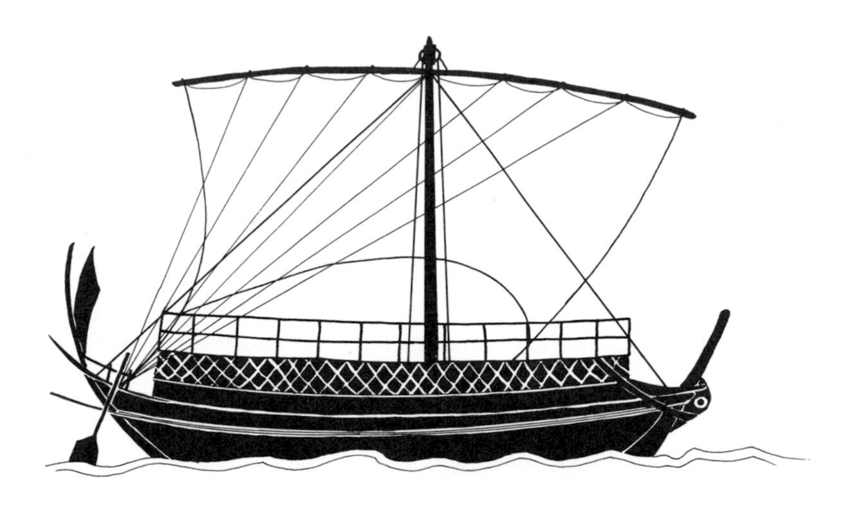

201 Reconstruction of a Greek merchant ship of about 500 B.C., based on a number of painted vases of the period

The penteconter, bireme and trireme were in essence men-of-war, although they may all have been used for trading in times of peace. The Greek merchant ship of 500 B.C. was, however, a very different vessel. It was clearly developed from the Cretan type of ship already discussed (Figure 130), and was probably only fractionally larger. It is often shown with a fiddle-bow. One finds few illustrations in which the cargo was clearly stowed on deck, and must assume, therefore, that the hull was more sturdily built than in the earlier Cretan ships, and that the cargo was carried largely in a hold.

other sea-going people of the Eastern Mediterranean, more notably the people of Syria and Egypt, had begun to build large ships with three banks of oars, the so-called triremes, and the Greeks, threatened with the possibility of an invasion by the Persians, for whom these other people of the Eastern Mediterranean acted as mercenary seamen, began also to build triremes. Much has been written on the subject of the trireme, but the truth is that we know very little about their construction or even the way in which they were rowed. The sole evidence upon which we must base our reconstruction of them lies in a few literary sources, some very fragmentary marble statuary and a single rather crude clay model, a few paintings and some sketches which are possibly, but not certainly, intended to depict triremes. What does seem certain, however, is that the total number of rowers varied from 120 to rather more than 200 and that they were arranged in three banks of which those in the middle bank sat further inboard than the others. Finally, the literary evidence shows us that they were far from stable and that when ramming, which was the principal means of attack, in order to steady the ship the blades of the oars had to be allowed to drag in the water. We frequently read, also, of ships being swamped in heavy seas and of rammed ships capsizing. Nevertheless, there can be no question of the length of the trireme being restricted by the use of a single timber for the keel-plate: the keel must have been composite. Although we have no knowledge of construction methods, the shipwrights were obviously now capable of building not only longer men-of-war but also mercantile shipping capable of carrying large cargoes.

Larger and better shipping meant of course that the Greeks were able to defeat the Persians at sea and remain a major sea power in the Eastern Mediterranean, although in the Western Mediterranean they were to come increasingly into conflict with Carthage and later still with Rome. But in the meantime the Greek and Persian wars brought the Greeks into far closer contact with the people of the Near East than they had ever been before, and from them they learnt little that did them any good, for their newly acquired wealth added impetus to a social change that had already started to take place. Even during the Persian wars Greek shipping was manned by freemen, but this was not a state of affairs that was to last for long and soon we read of slaves being used to row Greek shipping, and from then on the social structure of Greece became more and more similar to that of the Orient. Very soon all industrial enterprise was based upon slavery, a state of affairs that Greek philosophers bent over backwards attempting to justify.

7
The Engineers

(300 BC–500 AD)

The year 500 B.C. represents in antiquity almost the end of one aspect of technological development in the Near East and indeed in the whole of the Western World. It is true to say that virtually no new raw material was to be exploited for the next thousand years and that no really novel method of production was to be introduced. What new advances were made were to be almost entirely in the field of engineering, and most of the principles involved had themselves already been discovered and applied, although usually on a smaller scale. Thus, during the entire period in which Rome dominated the civilized Western World one of the few really novel means of production that appears to have been developed in any industry was that of blowing glass. Hitherto glass vessels had been made either by dipping a core into molten glass, the core later being removed, or by winding rods of heated softened glass round the core and finally rolling this shape backwards and forwards over a flat surface to smooth it. The process of glass-blowing had, in fact, to wait for the development of iron technology, for of the metals known in antiquity only an iron blow-tube would serve for the purpose of blowing glass. It was not until the beginning of our era that glass-blowing, therefore, became a widespread practice, but while the Romans may well have been responsible for the spread of this technique certainly no Roman ever invented it, for it seems to have developed either in present-day Syria or one of the countries adjacent to it.

Historians reviewing this state of affairs, this falling-off in the rate of technological innovation, have tended to attribute it to a number of causes, the first of which was the widespread use of slave labour. The dirty end of production, it is argued, was put entirely into the hands of slaves, and increased output could be achieved only by one of two means: either by acquiring more slaves or making those that one had work the harder. Since it is not in the nature of slaves either to invent new means of production or to exploit new materials, the possibility

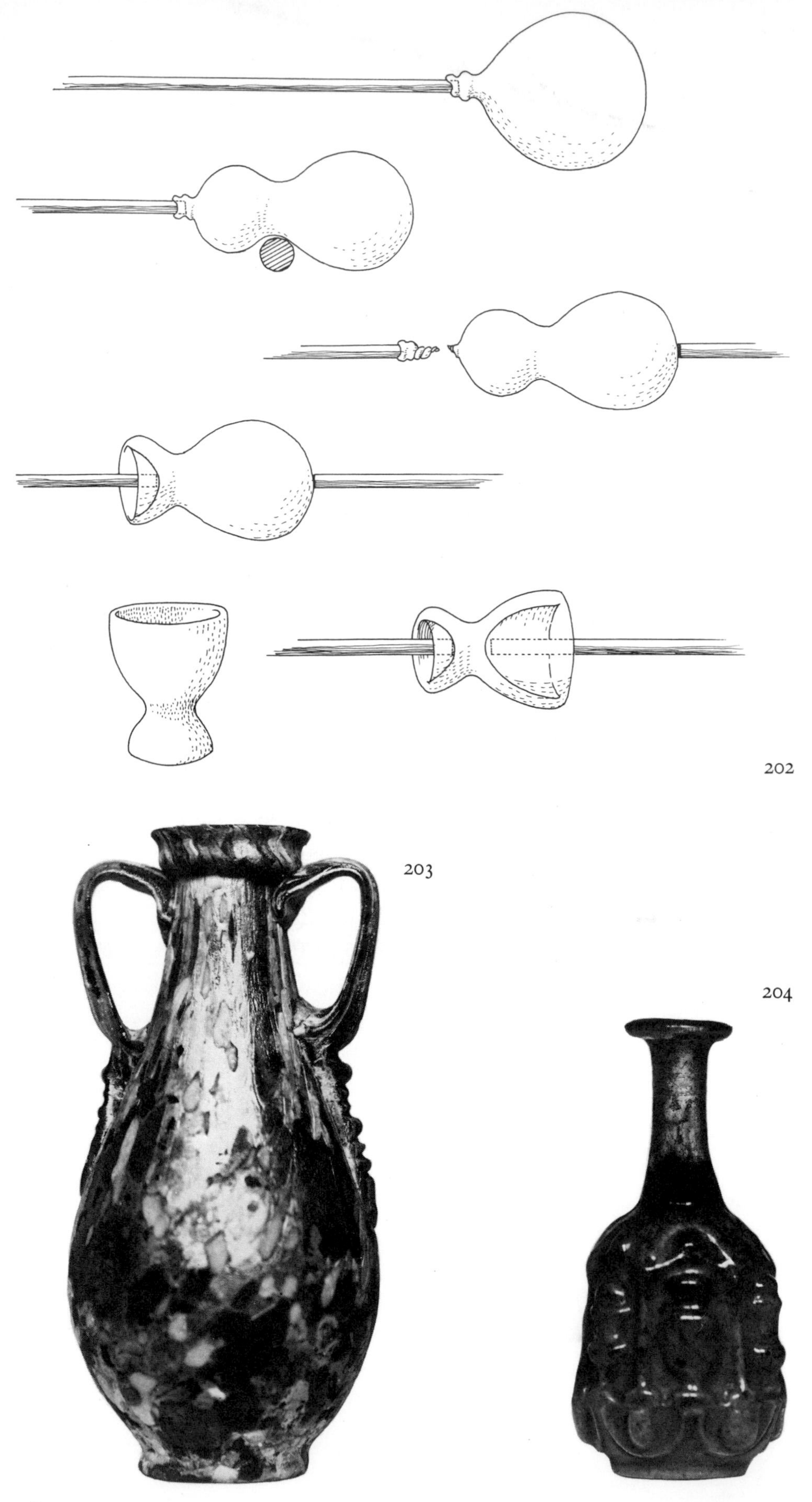
202

203

204

202 Diagram showing the stages by which a Palestinian 'wine-glass' was made in the first century B.C.

203 A 'free-blown' glass bottle of the first century A.D.

204 A 'mould-blown' glass bottle of the first century A.D.

The blowing of glass vessels, which was introduced in the second or first century B.C. in the Levant, was to replace almost completely the earlier method of moulding glassware. Either vessels were free-blown – the diagram shows the stages by which a plastic bubble of glass was converted into a 'wine-glass' – or the glass was blown into a piece-mould as, indeed, is most of the glassware made today.

of any further technological development came to an abrupt end. Alternatively, it has been argued that in the early centuries of the Empire, that is to say until the beginning of the decline of Roman power, there was an excess of manpower and that any improvement in production would merely have resulted in massive unemployment, with which the rulers of the various provinces would have been unable to cope, so that technological development was, therefore, actively discouraged.

A third contributory cause that historians have often ignored is the appearance of a very mediocre and often dishonest class of lower-grade administrators and civil servants into whose hands was put the running of industry, commerce and agriculture. Their business, as they interpreted it, was to see that the output of workshops and farms was adequate, and that any commercial enterprise such as shipping ran at an acceptable level of profit. These men were responsible for all the technological processes involved in the enterprises that they were controlling, yet their recipes for increasing output were seldom other than to employ more labour. Thus, in a large-scale process such as the cupellation of lead ores for the recovery of silver where, due to lead poisoning, the life of a slave could be reckoned in months rather than years, nothing was done to improve the unhappy lot of the slaves although the cost of replacing them was enormous and had to be borne against the profits of the process. It was this lower order of administrators whose duty it was to understand the processes of which they were in charge, and it was from their ranks that new technological innovations should have come. Instead they behaved like perfect civil servants: they kept their businesses, their estates and their workshops in good order; asked no intelligent questions and got, therefore, no intelligent answers.

When one reads what was written by philosophers and statesmen of this period there is much in their writing to support this point of view.

Indeed, the general gist of what passed for philosophy, insofar as industry was concerned, was that technology was a filthy business fit only for slaves and that no intelligent thinking man would bother his head with it. But for reasons that none of the philosophers are able to make quite clear it was apparently quite in order for a member of the upper crust to concern himself with engineering in its widest possible sense.

During the thousand years of which we are speaking, which embraced the height of both Greek and Roman power, there were in fact few respectable occupations. One could be a politician, a philosopher, a jurist or a general or, preferably, all four. It was also quite acceptable to be an artist, provided one had a sufficient number of slaves to do one's dirty work. Under such social conditions it is not difficult to see how the engineer came into his own. He could build a city water supply or drainage system and this would enhance the current politician's reputation. Alternatively, he could devise machinery essential to the better conduct of warfare which would make him acceptable to the military. To be blunt, because the results of the engineer's handiwork, which in itself was neither particularly arduous nor dirty, were immediately apparent, the field of engineering was, as we have said, the only aspect of technology to which an intelligent man could apply himself.

Having been forewarned that this chapter will have to deal largely with mechanical devices and with building we must now return to the situation in Greece. The Greeks, having been conquered by Philip of Macedon, were in turn carried to conquest in Asia and Egypt by his son Alexander. At his death in 323 B.C. Alexander had engulfed the whole of the civilized world with the exception of the small colonies at the western end of the Mediterranean. This vast empire was divided amongst his generals, one of whom, Ptolemy, took as his part Egypt, proclaiming himself king and giving himself the name Soter, or Saviour. Despite this exaggerated view of his own importance, Ptolemy established what was essentially a research institute by founding the Museum at Alexandria, of which in time the library was to become the most famous in the world. To the Museum came a great number of scholars both to learn and to teach, amongst whom one of the most important was Hero who had been a pupil of Strato, a contemporary of Aristotle at the Lyceum in Athens. Indeed it is difficult to say what proportion of Hero's writings depended upon the work of Strato and how much was original. He compiled, nevertheless, what could correctly be called a textbook of engineering. But not all the scholars at

205 Diagram showing the mechanism of Hero's water-clock, based on literary sources

206 Diagram showing the mechanism of Ctesibius' water-organ, based on literary sources

207 Diagram showing the mechanism of Ctesibius' fire-engine, based on literary sources

Most of the machines developed by the scholars of the Museum at Alexandria in the last two centuries before our era are known only from literary sources. The three hydraulic devices shown here are reconstructions. The need to graduate the drum showing the hour in Hero's clock was dictated by the fact that the period from dawn to dusk was subdivided into hours which, therefore, changed in duration with the seasons. Ctesibius' organ worked on the principle of displacing a volume of air by water, while his fire-engine – a double-action pump – depended upon the use of clapper valves which were probably already employed in bellows used by metal-workers.

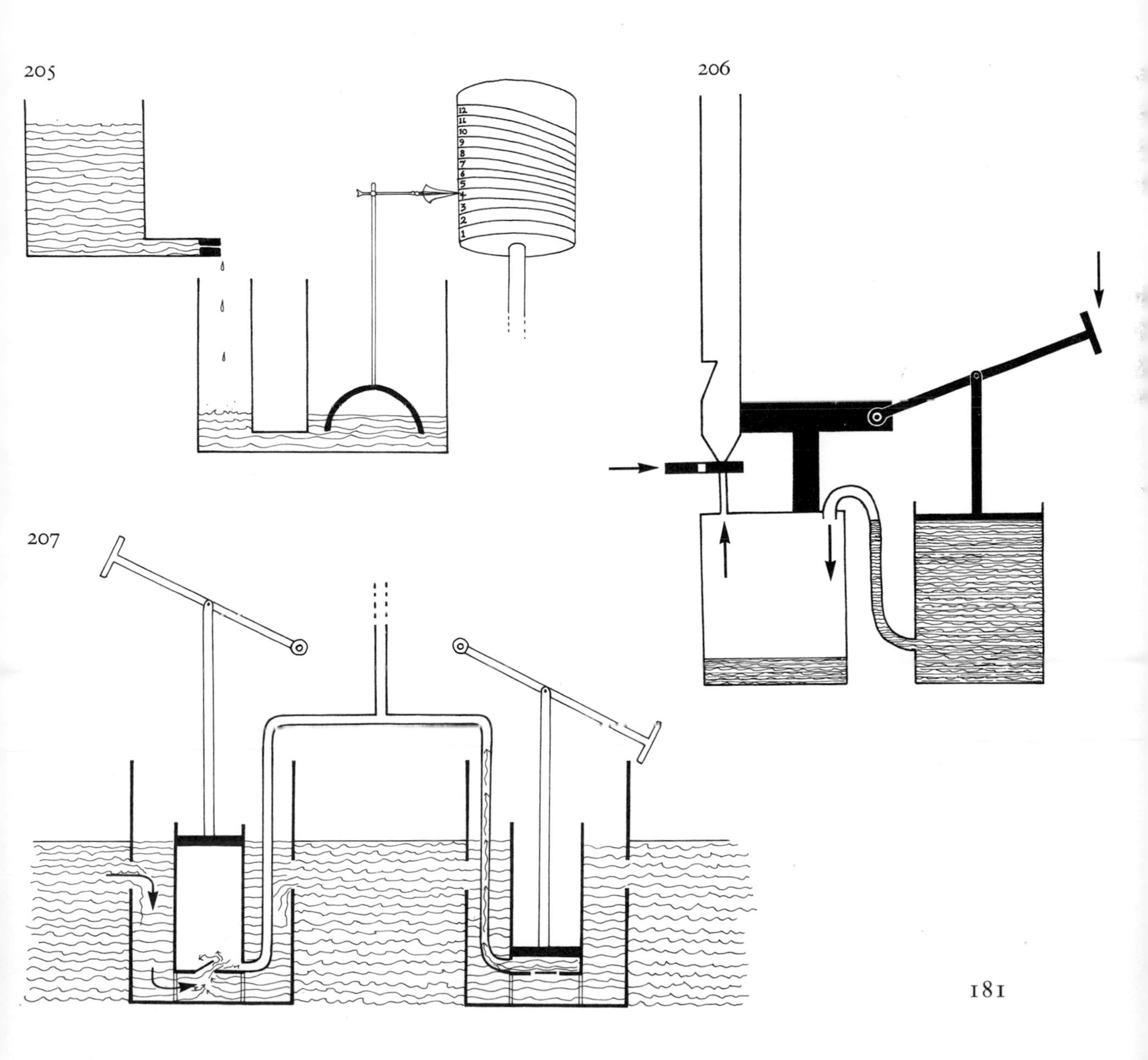

208 Diagram of Hero's device for opening temple doors

209 Hero's steam turbine, a diagram based on literary sources

Two devices which depended upon the expansion of air or the vaporization of water when heated. The mechanism for opening temple doors when a fire was lit on the altar was typical of the 'magical' purposes to which the inventions of the Museum were too often put. Hero's steam turbine could have been further developed to provide a useful source of power. As it was, the machine appears to have been regarded simply as an amusing toy.

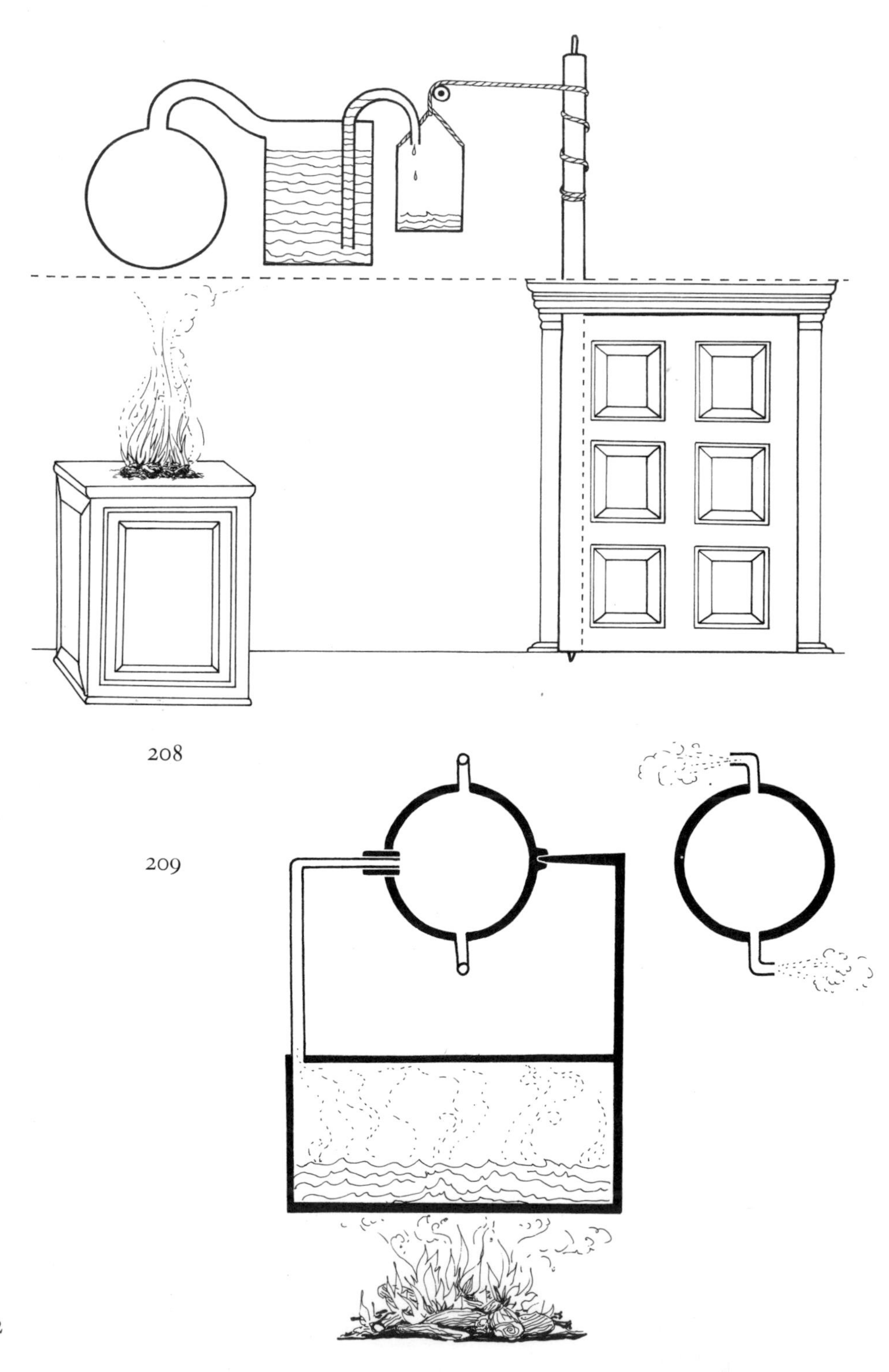

208

209

Alexandria were imports for there were local products also, as for example Ctesibius, the son of an Alexandrian barber. Hero, Ctesibius and another Greek contemporary, Philo of Byzantium, were responsible for the invention not only of a considerable number of useful pieces of equipment, but also of several interesting gadgets that could have been put to useful purposes, given the right social atmosphere.

Most of these inventions depended upon a few simple principles including the syphon, the spring, the screw, the pulley, the lever, cogs, cams, valves, and the fact that, when heated, air expands. Apart from machinery incorporating these elements great improvements were made in clocks, using water as the means of movement, and in astronomical instruments. Hero, for example, invented a water-clock and from his written description it is evident that he gave great thought to maintaining an even flow of water into the mechanism so that it would keep accurate time. Ctesibius occupied much of his time in devising pieces of artillery and is credited with at least one weapon that worked by compressed air. But, alas, the materials of the day were not up to the design and the machine never functioned properly. He had more luck with his fire-engine which was, in fact, a double-actioned pump. Philo also worked extensively on artillery and produced a treatise on ballistics in which he attempted to explain the forces at work when the machines were fired.

Much of the output of these inventors, however, went into creating interesting gimmicks often used in temples as, for example, doors that opened and closed when a fire was lit on the altar. The hot air from the fire caused a volume of air to expand and so, through a system of gears and pulleys, opened the doors. Ctesibius invented a water organ in which air was drawn through the pipes by a falling column of water. Hero even invented a primitive steam turbine, but it appears to have been put to no mechanical use. Indeed, when one looks at the writings of Philo, which include such topics as the defence and siege of towns, the building of harbours, the principles of the lever and pneumatics, and his work on ballistics, one can appreciate the restricted uses to which the brain-power of the Museum was put. Essentially it was for one of three purposes – to beautify the cities, to serve the army or to mystify the worshippers in the temples. At no time does it seem to have crossed the minds of these ingenious men that their inventions could have been used to provide new sources of power or to make industry more efficient, and yet they came within a hair's breadth of a real industrial revolution. Many of Hero's inventions, for example, could have been applied to the production of power. He even devised a

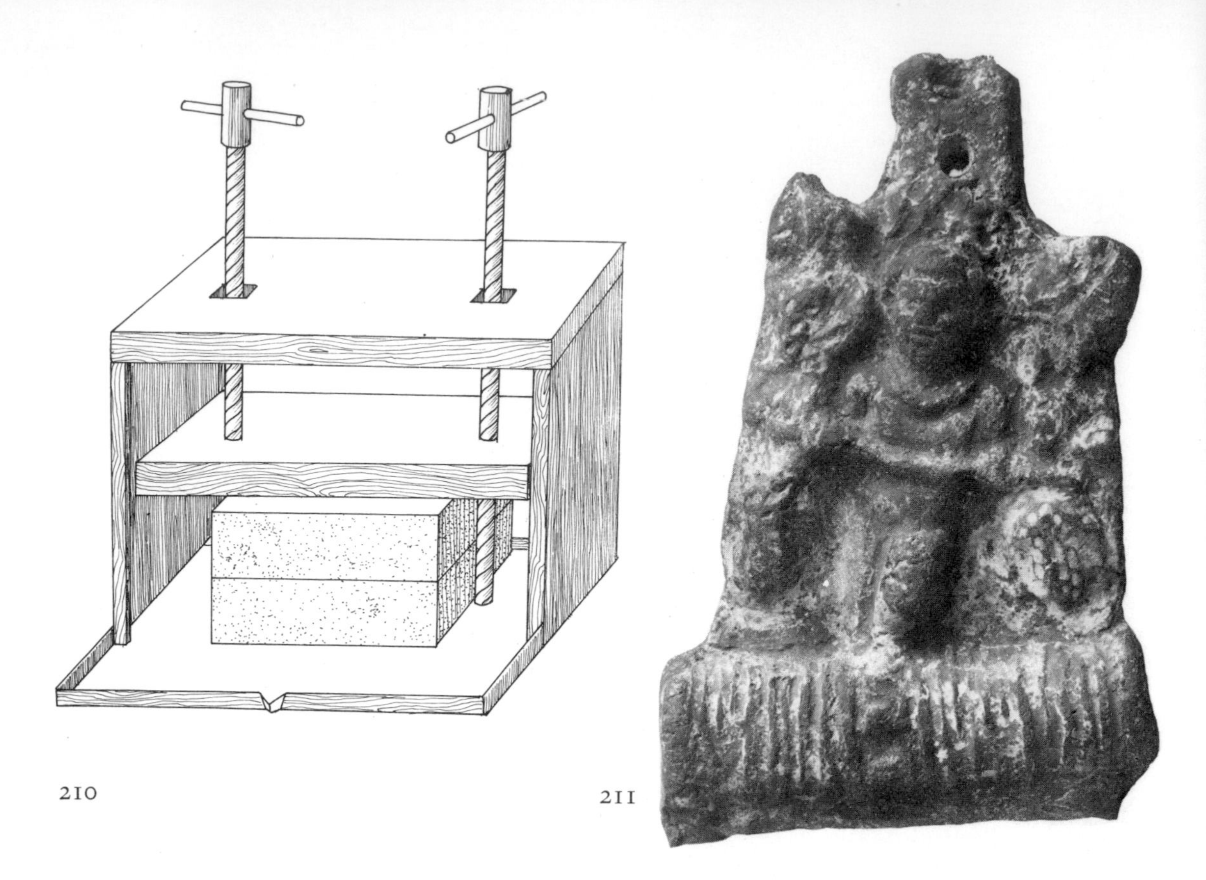

210

211

screw-cutter which would have allowed the assembly of really sound machinery. But the failure of the Museum is, perhaps, summed up best in the piece of artillery worked by compressed air invented by Ctesibius. The inventiveness was there, but it was applied to a non-productive purpose, while the materials available and the means of shaping them were inadequate for the manufacture of so sophisticated a piece of equipment.

The final word, however, must go to Archimedes – undoubtedly the greatest engineer of this age. Born in Syracuse, he is said as a young man to have visited the Museum. His inventions included a screw-pump for lifting water from mines, systems of compound pulleys and levers for raising heavy objects, as well as a great number of military engines for the defence of his native city, amongst which was the proposed use of large concave mirrors in order to concentrate the sun's rays upon attacking ships to set them afire. As with the others at Alexandria, however, few of his inventions could have been as efficient in practice as they should have been in theory; and yet he gave no thought to improving the materials of which they were made or the manner of their construction. Indeed, when he was asked if he would write a handbook on engineering he refused, giving as his reason that the work of an engineer and, indeed, everything that would in any way make life easier, was ignoble and vulgar.

210 Diagram showing the screw-press described by Pliny

211 Terra-cotta model of a slave working an Archimedes' screw-pump

212 A screw-pump in use in Egypt today

213 Diagram showing the way in which the screw-pump functioned

A small proportion of the machines developed by the scholars of the Museum and by Archimedes depended upon the use of the screw. It is difficult today to say when and where the screw was first used. It may well be that the screw-pump, to which Archimedes' name has been given, was already in use in Egypt before his time, and that he merely 'popularized' the machine. In one of his treatises Hero describes a device for cutting screws. The screw-press for extracting oil from olives may be seen as typical of the purposes to which the screw was most widely applied.

212

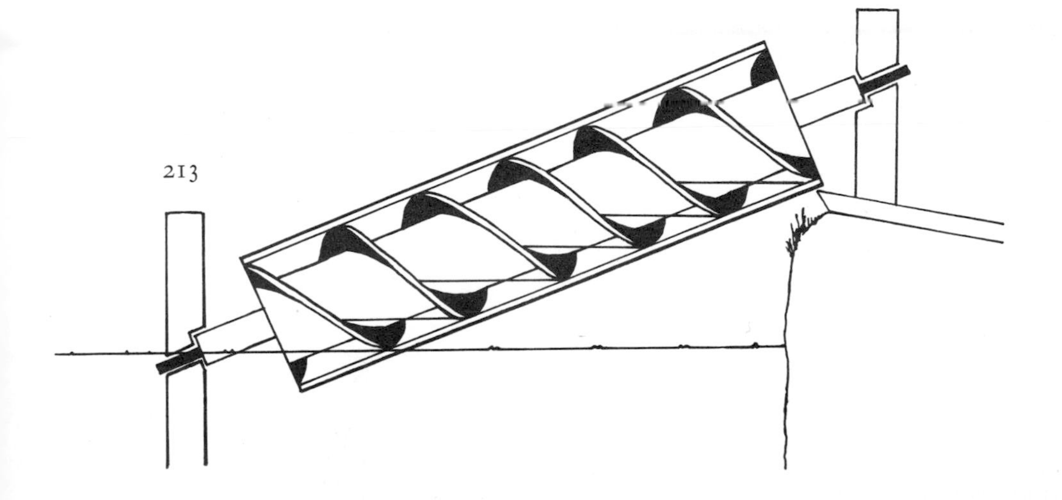

213

214 Reconstructions of the catapult (a) and cross-bow (b), based on literary sources, and a photograph of a working model catapult manufactured by a modern firm of toy-makers

A considerable part of the work of the Museum scholars was devoted to the design and study of military equipment. The cross-bow and the catapult, in which the power was provided by skeins of sinews held under tension, were both later to be improved upon by the Romans. It is far from clear from written sources how the cocking and trigger mechanisms worked in these weapons. A number of treatises on ballistics show that the students at Alexandria were concerned with increasing both the accuracy and the range of these devices.

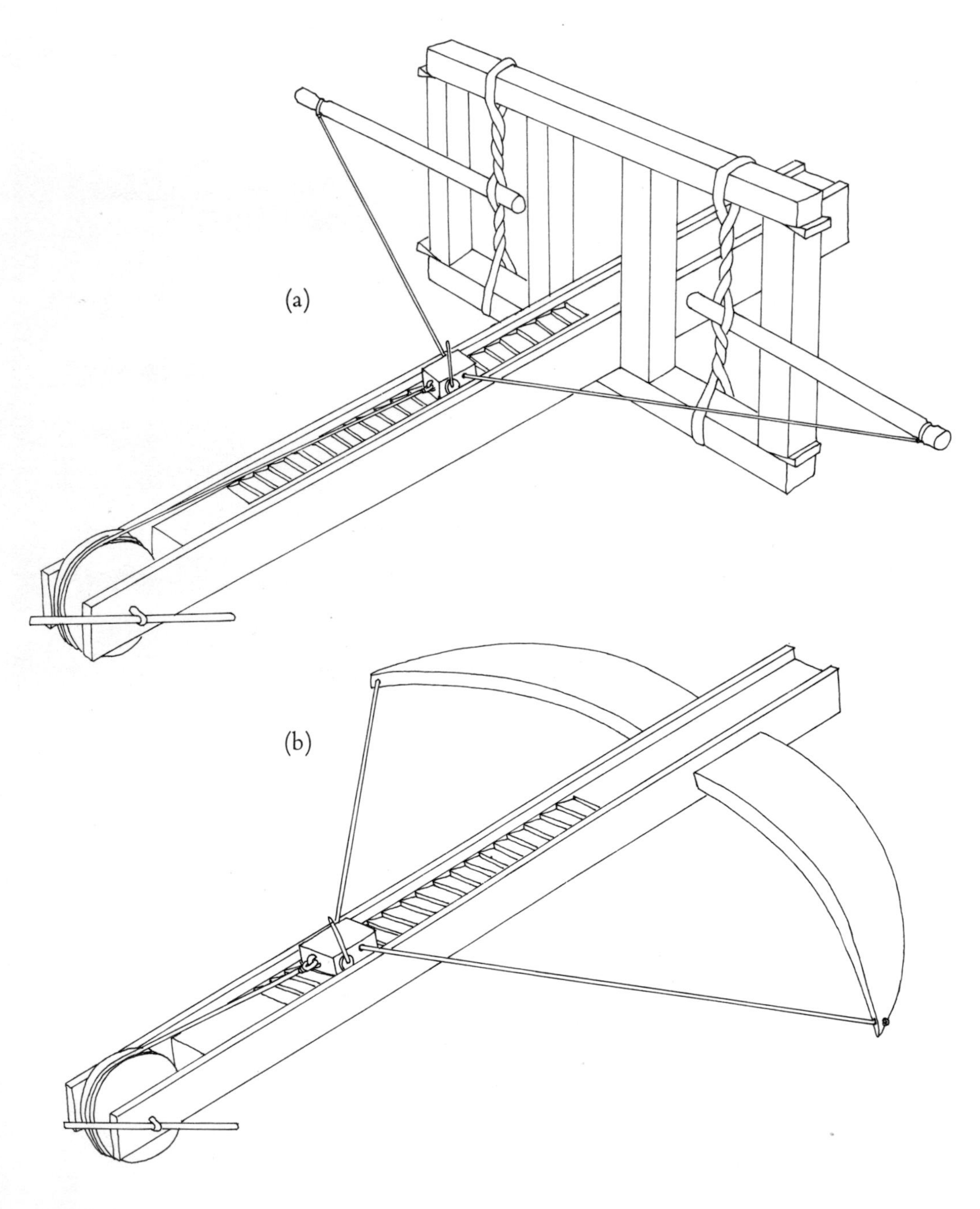

It would be totally wrong, however, to imagine that the inventions of these great men were put to no practical purpose at all. From this time on cogwheels and pulleys, levers and pumps were widely used throughout the ancient world, and their application certainly simplified a great number of processes, more particularly that of building, where the lifting of heavy stones, for example, became far less of a problem. When we look at the illustrations of machinery of this age what is so surprising is not so much that it existed at all, for there were plenty of minds to create it, but that it was so clumsy in construction and that the materials used were often so inadequate for the function to which they were being put, while to overcome the inadequacies of the materials being used machinery often had an enormous bulk which must frequently have been a trial of strength to move and manipulate.

Unlike Archimedes, the Romans do not seem to have cared unduly whether engineering was vulgar or not, and in their rapid conquest of the civilized world they soon learnt to appreciate and to copy other people's mechanical devices. Indeed, the major contribution of Rome to the advance of technology lay in the ability of its citizens to absorb ideas from elsewhere and to provide an administration that would allow them to be used to their greatest advantage. The layman is probably familiar with many aspects of engineering as practised by

215

215 Roman relief showing a crane in use during building operations

216 Diagram showing the probable elements of the Roman crane

The Greek scholars also devoted a great deal of attention to the study of systems involving pulleys. Tackles for lifting heavy weights were obviously much in demand for the construction of large buildings, as in the case of the Roman crane depicted here. As a means of providing power the tread-wheel had already been applied to the screw-pump, although in a very much smaller form (see Figure 211).

Rome, and yet possibly fails to appreciate that virtually every aspect had been invented and practised elsewhere often long before it was acquired by the Romans. The large public buildings of Rome were inspired by, if not actually designed by, Greeks, and although the Romans appear to have excelled in the art of building aqueducts (in the first century A.D. Rome could count nine such structures bringing water to the city) the Greeks, Assyrians, Babylonians, Persians, and Egyptians had all been building aqueducts for many centuries before. The same may be said of the city's drainage system. The Roman roads which crossed the empire from boundary to boundary were often no better than the Greek and Persian roads that had preceded them. Indeed one author has rather scathingly said that the Roman concept of a road was that of a wall buried in the ground, the top of which was paved and along which the traffic moved, a view which, though somewhat exaggerated, embodies a little truth, for Roman roadways were often elaborately heavy when one reviews the purpose to which

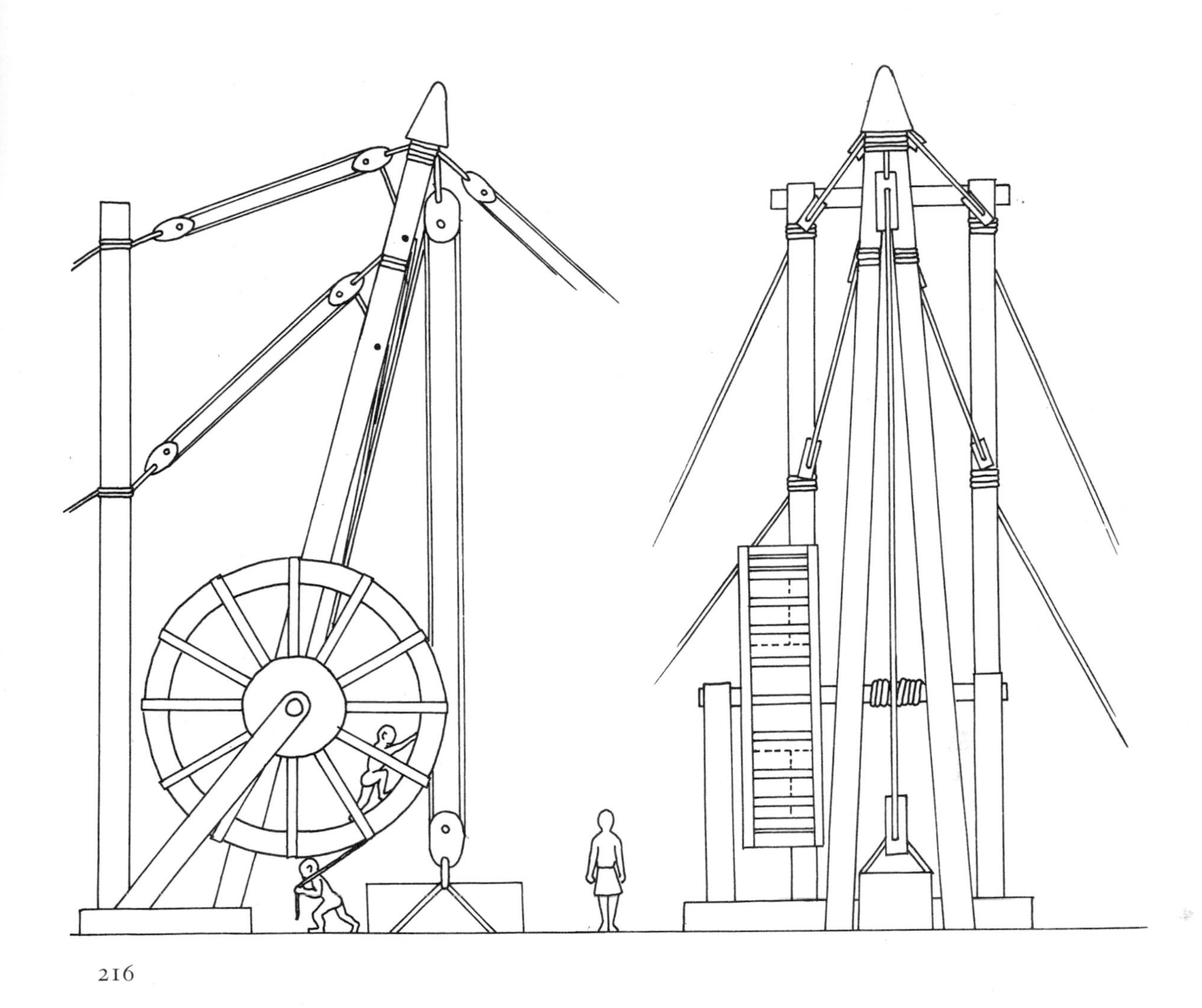

216

they were intended to be put. Equally the greater part of Rome's military equipment, particularly machinery such as the catapult, as we have just seen, was of Greek invention, while her early naval vessels were timber by timber copies of those used by the Greeks and Carthaginians.

It would, however, be quite unfair to give the impression that the Romans did not appreciate technology even if they were not particularly inventive. Nowhere is the appreciation of what could be achieved by applied technology more clearly seen than in the writings of the Latin encyclopedists. Strabo, for example, who died in about 25 A.D., filled his compendious *Geography* with comments on technological processes, and although many of his observations were at fault in detail, at least he felt them worth recording. The same can be said of Pliny, who was born just before the death of Strabo, for a large part of his *Natural History* is occupied by a discussion of raw materials and the techniques used to work them. The Roman attitude to technology was, indeed, subtly different from that of the Greeks. The Romans, like the Greeks, still believed that actually to work with raw materials, to be an artisan, was degrading; but nevertheless they did not hold technology in such disdain that they refrained from writing about it. Thus for example the Greek philosophers were apt to write about medicine and surgery as abstract studies, while the Romans not only followed them in this but also built hospitals and organized a medical service for their army. To the Greeks mathematics were on the whole an interesting mental exercise, but it was the Romans who used mathematics to build roads and aqueducts and to lay out their somewhat stereotyped city streets. It is to this aspect of Roman genius that we must devote the last pages of this chapter.

One of the most interesting features of the technology of this period was the use made of the mechanical devices that Rome had inherited. These were all based essentially on gears, pulleys, the screw and the lever. Thus we find that the crane used during the erection of large buildings was essentially a sheer-leg tripod, the load being raised by means of compound pulleys, a system that probably differed in no way from that used by the Greeks except that as motive power the Romans attached to it an enormous tread-wheel, which itself was possibly developed from the same type of device used when raising water with a screw-pump. Elsewhere, instead of the tread-wheel we find the capstan was being used, which in turn was borrowed as a concept from rotary olive and wine presses or from corn mills. By the end of the fourth century A.D. an unknown writer speaking on the subject of warfare goes so far as to describe a paddle-driven ship in which the paddles were driven through gears by capstans on the deck turned by yoked oxen. Probably such a ship was never actually built, but it is interesting to see even the theoretical appearance of a paddle-wheel at this early period.

Mention of the paddle-wheel brings us to one of the more curious aspects of Roman engineering, namely the spread of the water-mill throughout the Roman Empire. One does not know when and where the first water-mills were developed, although written evidence shows that they were certainly functioning by the first century B.C. in Northern Greece, and it was presumably either here, or in Western Anatolia, that they were first put to use. They were, to be precise, water-turbines, and they could operate only in an area where fast streams could be channelled to produce a jet of water. The wheel was set horizontally, that is to say with a vertical axis, and the water was directed on to the blades of the turbine. The axle passed through the

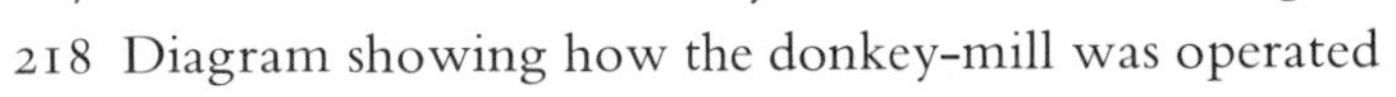

217 The lower stones of donkey-mills, from a building at Ostia
218 Diagram showing how the donkey-mill was operated

217

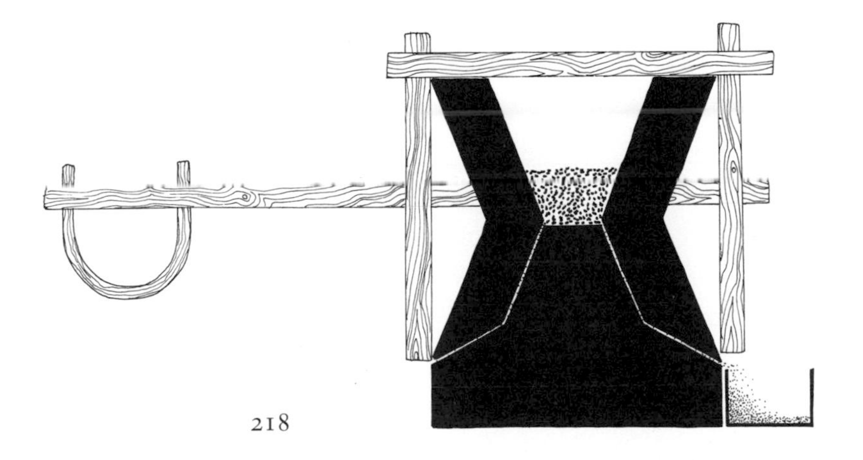

218

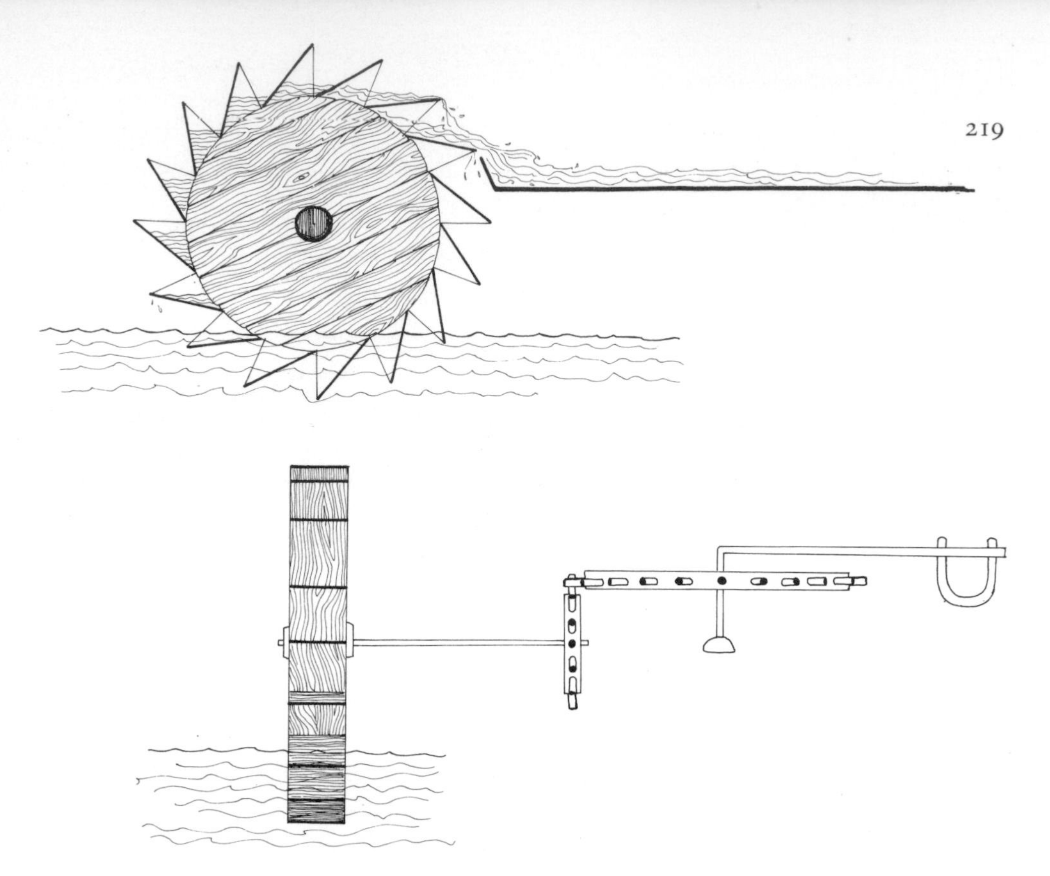

lower horizontal millstone to rotate the upper stone, there being no gearing, so that the speed of the turbine was the same as that at which the upper stone turned. Mills of this type are still to be found in many parts of the world today, but only in those areas where the river system will allow the water to be run through narrow channels and ultimately down a steep shute directed on to the blades of the turbine. In lowland areas where streams are sluggish such a mill could not be made to work.

Quite a different type of water-mill first described by Vitruvius in the first century A.D. worked on the principle of the water flowing beneath the wheel, which was set vertically, and in so doing striking the blades to make it rotate. The water-wheel drove the millstones through a gearing in which the ratio was five turns of the millstone to one of that of the water-wheel. Such a water-mill as this obviously had a wider application than the previous type, and furthermore it was capable of doing far more work. Whether or not it was developed from the horizontal water-wheel is a matter of debate, but it should be noticed that it bears a strong resemblance in design to the water-raising wheel in which a draught animal working through a capstan rotated a wheel bearing cups at its rim, each cup dipping into the river and being discharged into a shute after little more than half a turn of the wheel. There is reason to believe that water-raising wheels of this sort had been in use in Egypt for some centuries before our era and it may well be that the so-called Vitruvian water-wheel was developed from these.

219 Diagram showing how the water-raising wheel operated

220 A water-raising wheel in use in Cyprus today

221 Diagram of a paddle-driven ship, based on literary sources, but probably never built

During this period the capstan appeared in many guises. In its simplest form it is to be seen in the rotary mill in which the upper stone was made to rotate by a horizontal beam to which was yoked a donkey or an ox. With the addition of simple gearing the same principle was applied to the water-raising wheel, first known from Egypt and still to be seen in use there today. The rather complex paddle-driven ship, in which the paddle was worked through gears by ox-drawn capstans on deck, is known only from literary sources, and in all probability was never put to serious use.

220

221

222 A recent picture of a horizontal mill from Norway

223 Diagram showing how the horizontal mill operated

Mechanized milling of corn is known from written sources to have been developed by the first century B.C. in Northern Greece and Western Anatolia. The type of mill used appears to have been essentially a turbine in which water from fast mountain streams was directed through a chute on to the blades of the water-wheel, so rotating the upper millstone. This device seems to have spread rapidly throughout the Roman Empire, and in basically the same form is to be seen in many mountainous areas today from the Middle East to Northern Europe.

222

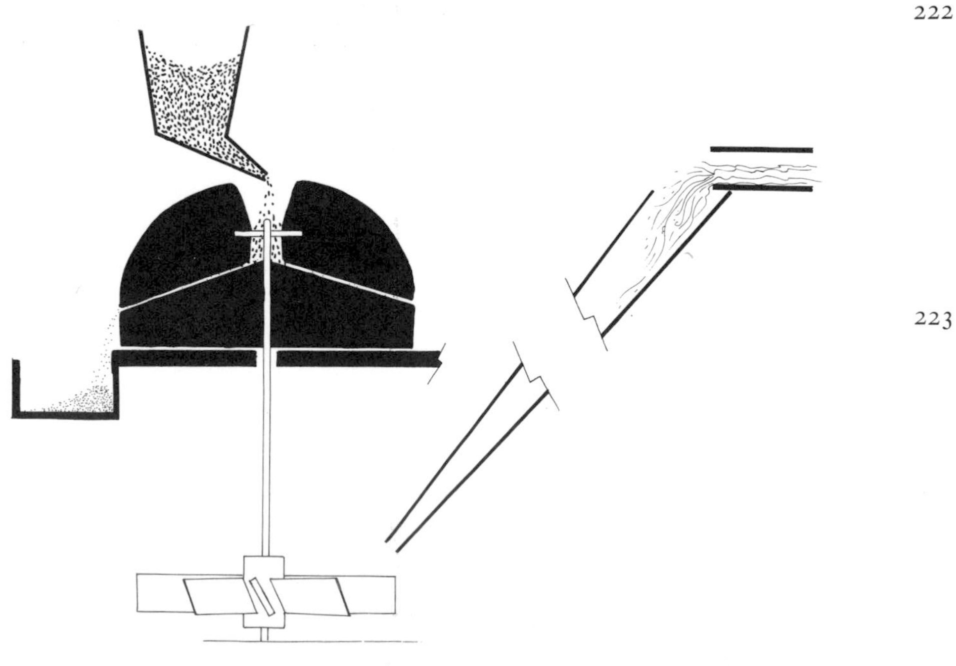

223

224 Drawing of a mosaic from Byzantium, partly reconstructed, showing a water-mill of the Vitruvian type

225 Diagram showing how the Vitruvian mill operated

226 Diagram showing the functioning of an overshot mill

A water-mill that could be made to work in regions where rivers were sluggish was first described by Vitruvius in the first century A.D. This was the 'undershot' mill in which the water passed below the wheel, striking the blades to make it rotate. The millstone was made to work through a gear system very similar to that used in the water-raising wheel previously described. It is apparently this type of mill that is illustrated in a mosaic from Byzantium, modern Istanbul.

The Vitruvian mill could operate only when the water in the river was neither too high nor too low. With the later introduction of the 'overshot' wheel, in which the water was ponded back in a pool kept at constant level, water-mills could be made to operate in nearly all regions of the Roman Empire.

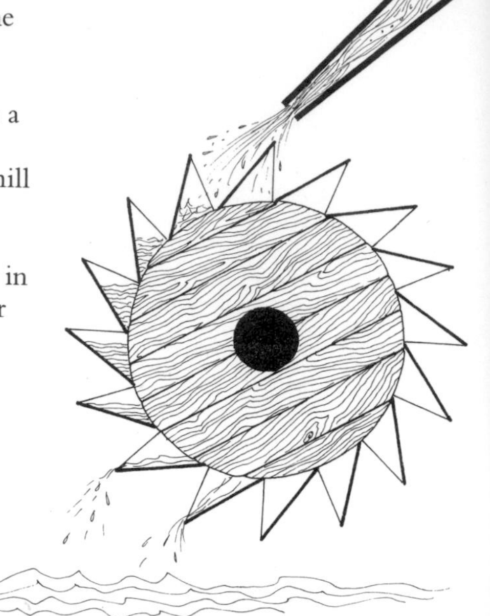

Be this as it may, the Vitruvian water-wheel had its limitations, for it was dependent for its proper functioning on a level flow of water in the river: were the river in spate or too low the wheel would not function.

By the fifth century A.D. yet a third type of wheel appeared. This was the over-shot wheel in which the water, ponded back in a mill-pool, was directed by means of a shute over the top of the water-wheel, and which thus, by controlling the level of the water in the mill-pond by means of sluices, could be made to function at most seasons of the year throughout the larger part of the Roman Empire.

Curiously enough the Romans seem to have used the water-wheel largely for grinding corn or for similar purposes, such as extracting oil from olives, although at times great ingenuity was shown in devising mills. Thus, there is one recorded instance where a row of wheels was set one above another up an incline so allowing the same volume of water to turn a whole battery of mills. Equally, when Rome was besieged by the Goths in the early part of the sixth century a floating mill was moored across the Tiber, the flow of the river turning the undershot wheel. But at no time do the Romans appear to have adapted the wheel to any other type of process. There was, for example, no mechanical reason why the water-wheel should not have been used to drive bellows or heavy hammers as was done in contemporary China or during the medieval period in Europe.

The Roman city is interesting more for its scale than for the novelty of its design. The aqueducts, the lead delivery pipes, and the earthenware piping used for drainage, had all been anticipated elsewhere, although the sheer number and length of Roman aqueducts is in itself impressive. Probably the Roman's greatest contribution to building was the development of the brick and concrete arch. This was essentially a brick arch reinforced with a heavy infill of concrete. The arch, once the concrete was set, was essentially a vast lintel and thus it exerted very little lateral pressure so that the pillars on which it was set required no buttresses to support them. It was a series of such archways that carried the larger part of the raised portions of the aqueducts, many of which still stand to this day. Curiously, for many of their early public buildings the Romans depended upon Greek architects, who introduced a style based almost exclusively on upright pillars capped by lintels, a somewhat cumbersome type of building which, despite its great aesthetic merits, was nevertheless designed to withstand earthquake tremors. However, apart from Rome and the major cities of the Empire, the more typically Roman building materials were used and very high technical standards were achieved not only in the minor provincial cities but also in the homes and estates of Roman administrators and businessmen. Thus, in Western Europe one need not be surprised to find well-constructed fortifications, city walls and public buildings within the cities. But the high quality of building in the Roman villas, for example, is quite extraordinary and shows that a great deal of money must have gone into their creation.

A number of Roman writers, in fact, deplored the building of these expensive estates, not only in Italy but throughout the whole Empire, arguing, probably quite rightly, that their construction was drawing a lot of wealth away from Rome itself, and we can see in this tendency yet one further cause for the lack of technological development during this period. Large country estates, no matter how well run or how well built, tended to be isolated and their occupants to become self-supporting. Thus, a well-set-up Roman estate would have its own farms and mills; its own ironworks and possibly its own pottery. Indeed the larger the estate became, the more likely it was to become self-supporting. The technological processes carried out on these estates would nevertheless have been on a comparatively small scale, and provided they supplied the immediate needs of the estate there would have been no impetus to increase their output. Just occasionally, however, we get a glimpse of some unusual development where, perhaps, there was a shortage of manpower. Such an instance may be seen in

227 Diagram showing the stresses involved in a true arch, the Greek pillar and lintel, and the Roman arch

228 A reconstructed view of aqueducts outside Rome carried on long series of arches

229 A view of the interior of a public lavatory, Ostia

One of the few original contributions of Rome to building technology was the introduction of cement. Apart from its use as a bonding material, it was also employed in the making of concrete which, combined with a brick facing, allowed the construction of solid arches, so eliminating the need for buttressing the supports. Arches of this type were to be seen in their most dramatic form in the building of aqueducts. Indeed, the Romans devoted a great deal of time and money to public sanitation, creating a water-supply, drainage and other forms of sanitation often as good as, if not better than, those to be found in many parts of Europe today.

227

228

229

the development of a reaping machine in Gaul which was described by Pliny. This was essentially a two-wheeled cart pushed by a pair of oxen yoked behind the vehicle. Along the leading edge of the cart was set a sharp steel comb at a height a little below that of the heads of the grain. The stalks of corn were cut by the comb and the heads fell behind into the cart. The machine was probably never very efficient, but when the period during which reaping could take place was short, and the work could not possibly be undertaken by the available number of men equipped with sickles or scythes, the device was obviously of enormous advantage. The machine could probably only be used in fields that were sufficiently large and level, but it is again symptomatic of the whole of the Roman period that it was not developed to become more efficient and did not achieve more widespread use.

Apart from the development of her luxurious roads, little progress was made by Rome in the field of transport, and although we know of a number of light vehicles in which the draught animals were harnessed between a pair of shafts and in which the normal yoke was replaced by a padded collar, for moving heavy burdens the older paired draught

230 Low relief from Belgium showing a reaping machine (vallus) at work

231 Reconstruction of this type of reaping machine, based on this and similar reliefs and on literary sources

In many respects the Romans were slow to appreciate the possibilities of mechanical devices. A simple reaping machine was developed in Gaul, for example, but it was neither developed further nor did it spread to other parts of the Empire, despite the fact that it could have greatly eased the shortage of man-power that crippled the later centuries of Roman rule.

230

231

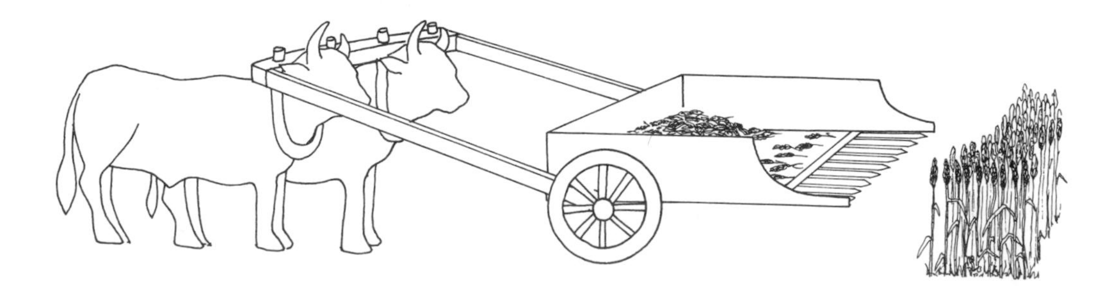

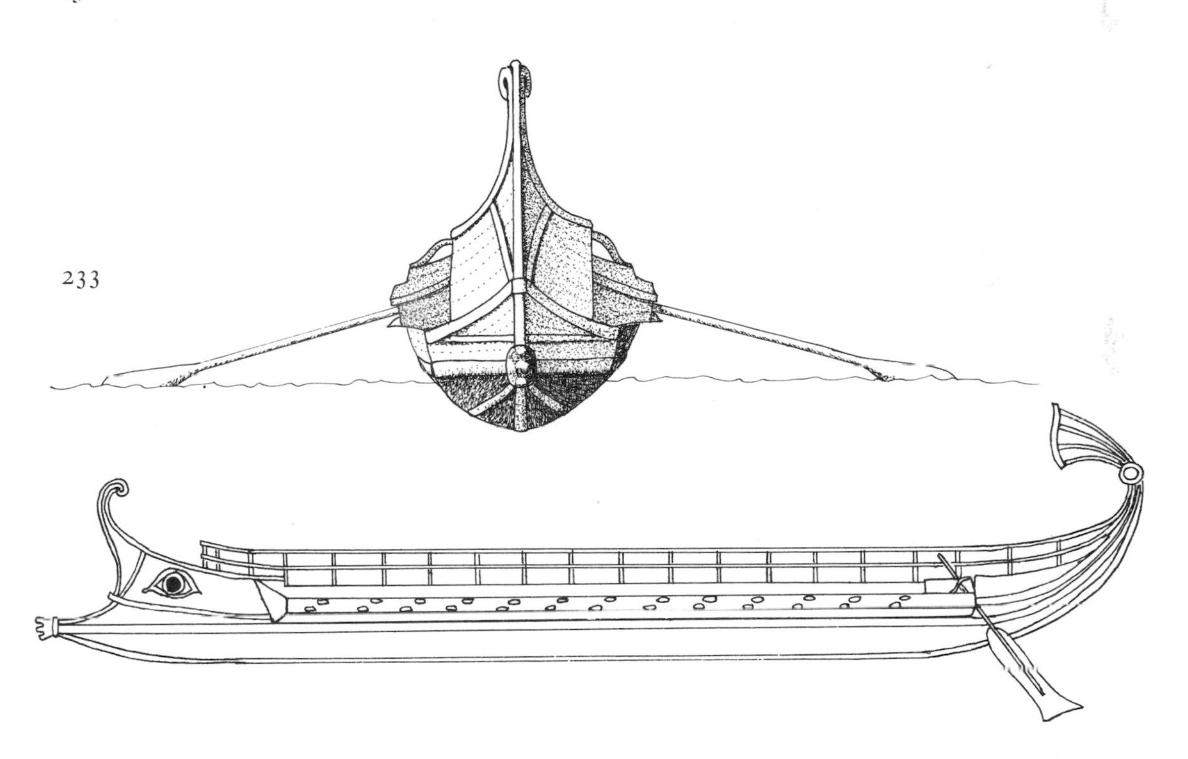

232 A Roman galley as depicted on a mosaic from Palestrina (Praeneste)

233 Reconstruction of the bow and profile of a Greek ship of about 200 B.C., showing the decked outriggers projecting beyond the ship's sides, based upon a badly damaged marble statue, vases and low reliefs

234 Roman merchant ships as shown on a mosaic from Ostia

Few changes were made in the design or construction of shipping under Rome. The war galley remained much as the Romans found it in Greece and Carthage. The three-bank ship seems to have been abandoned in favour of a two-tier system in which more than one man was set to each oar. Rowing was invariably carried out through decked outriggers, so protecting the crews and bringing the blades of all the oars into a straight line. Outriggers were certainly in use in Greece by the beginning of the third century B.C.

Merchant ships were often fitted with small square jib-sails, but otherwise the hull and general upperworks were very similar to those in Greece in the fourth century B.C.

and yoke were still used. Although the Romans were well aware of the potentialities of proper stock-breeding, they appear to have ignored the possibility of rearing a line of heavy draught horses, just as they failed to develop shafts and collars for the animals pulling their heavier wagons.

The same static position is to be seen in Roman sea and river transport. So far as their war galleys were concerned they borrowed from the Greeks and Carthaginians the low vessel with a ram, the oars now being rowed through decked outriggers which had the advantage of giving the crew a greater leverage while bringing the oars into a straight line, while at the same time the decking gave the oarsmen some protection from the enemies' missiles. Even by the fifth century A.D. Roman merchant shipping was much as it had been inherited from the Greeks: fairly short vessels high at both bow and stern. At some point, probably in the first century B.C., a small square jib-sail was adopted in addition to the square mainsail, but the vessels could still only be worked in a following wind. This severely limited the possibilities of sea commerce, since the cost of rowing merchant vessels for anything other than short distances was financially prohibitive. In the field of shipping, however, the lack of progress during the period of Roman domination is perfectly intelligible, for Rome had no competitors on the seas in which she did business other than pirates, whom her men-of-war were always capable of destroying.

Perhaps the most remarkable feature of the Roman period is the degree of sheer sophistication accompanied by utter crudity in much of the technology, even in essentially the same fields. Thus, for example, the Romans invented the most elaborate heating systems, not only the type of under-floor heating, the hypocaust, well known to most people, but also boilers with plumbed hot water available on tap; while their fundamental grasp of insulation allowed them to construct a primitive form of Thermos flask. Yet when it came to lighting, the lamp, which was no more than a small covered bowl with a circular hole through which protruded the wick, was barely any advance on that used by the artists who painted animals during the Ice Age deep in the caves of Southern France. Indeed if one wished to be thoroughly cynical about the matter one could say that the only advance made was in the cover over the lamp, which was not infrequently decorated with obscene reliefs. Again, we know from written sources that the Roman ideas of anatomy, physiology and medicine were extremely unscientific and that the proper function of many parts of the body had never been discovered; yet surviving surgical instruments show us not only that the Romans were able to carry out a number of reasonably complex operations, but also that their instruments were extremely well designed for the job. Many of their obstetric instruments, for example, can be paralleled almost in detail with those in use at the beginning of this century.

When the end came the Roman Empire was overrun by Barbarian tribesmen from Northern Europe and Nether Asia. In this final defeat some writers have seen a technological failure on the part of Rome: a failure in that she was unable to meet, despite her higher technology, the threat of these barbaric people. Some historians have gone as far as to say that these tribal horsemen, now equipped with the stirrup, were such an effective cavalry that the Roman legions were quite unable to withstand them. This view fails to take into consideration the fact that

235

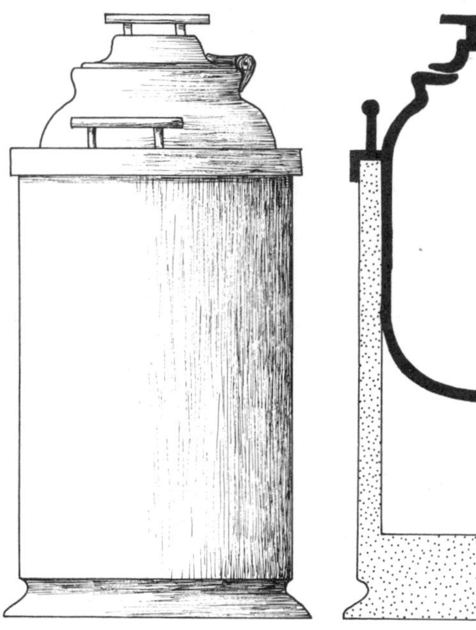

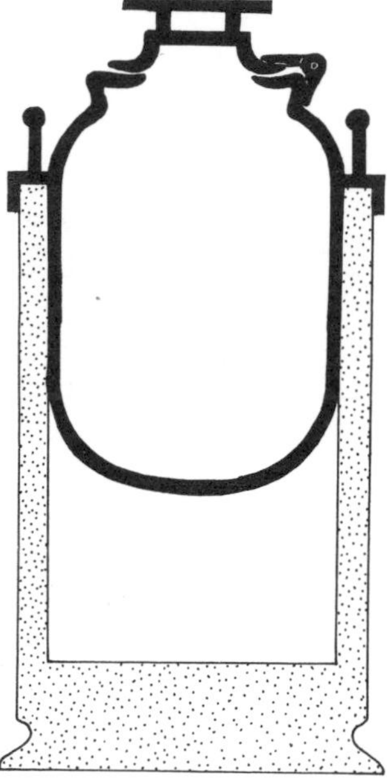

236

237

235 Copies of Roman and Old Stone Age lamps

236 A Roman 'vacuum' flask from Pompeii made of bronze and iron

237 A Roman 'air-bed': a woodcut of the sixteenth century based on Roman literary sources

Roman technology was riddled with strange contrasts. Their engineers were quite capable of installing plumbed hot water or of devising a 'vacuum' flask to keep food warm. On the other hand the normal form of lighting was a small oil lamp, a refined version, but functionally no different from lamps used in the caves of Southern France during the last Ice Age.

The inflated mattress, made of leather, was never widely adopted, if indeed ever used at all. This was a brilliant idea, but as with so many inventions of the period the materials of which it was made were hardly suitable.

for nearly three centuries the Romans had recruited many of their troops from amongst the Barbarians and that both Barbarians and Romans were well aware of the equipment and methods of fighting adopted by each other. Equally it fails to take into account the fact that well-mounted and well-equipped cavalry in the middle of the last century in the United States was not infrequently out-matched and utterly routed by poorly equipped tribesmen riding with neither saddle nor stirrup.

Before we pass final judgement on this question we should examine in rather more detail the state of technological advance of the Barbarian world.

8
The Barbarians

The Barbarians in the West

To the writers of classical antiquity, anyone who did not share in one of the great civilizations of the Mediterranean world was a Barbarian. The impression that they would have us gain is that these people were uncouth, untutored and thoroughly backward in every respect. And yet an author such as Caesar could barely conceal his surprise and indeed his admiration for the Gauls in the sea-worthiness and sophistication of the ships they sailed in the Western Sea. He was forced to admit, for example, that their ships were more sturdily constructed and better designed to withstand the weather of the Atlantic than those of the Romans. He even noted that their anchors were secured by chains and not by cordage. In effect what Caesar said was that these people might have been Barbarians but their shipping was excellent. The same writer, however, when he spoke of Britain, claimed that the inhabitants of the interior did not grow corn and dressed themselves in skins. But then Caesar had never set foot in the centre of Britain to see for himself what the situation was, and like so many authors of his day he was as much a dupe to propaganda as he was himself a propagandist. Sadly we must admit that the writers of classical antiquity are not the most reliable sources if we wish to know the state of technological advance of the Barbarian world.

This attitude of a higher civilization to a lower is of course still with us today. 'No peasant ever invented anything' is the rash statement of one eminent anthropologist. Even on the face of it this is an absurd statement, although it seems to embody an element of truth. One can ignore the statement at face value and assume that what is really meant is that peasants are less inventive than city-dwellers. We might do well, therefore, to begin by deciding whether the Barbarians were really less inventive than their more civilized neighbours to the south and east.

We have already spoken of the shipping of the Gauls which was so sturdily built that the Roman warships found it almost impossible to sink by ramming. These vessels, which Caesar used extensively later on as transports, were hardly the product of an unimaginative people. Equally, while Roman ox-carts creaked their way along their sophisticated roads, a simple carpenter in Northern Europe who had probably never seen a Roman, and who was certainly illiterate, designed and built a wheel that moved on roller bearings. In this context it matters little that his invention probably died with him and that he chose the wrong material of which to make this clever device. Yet another peasant, this time living in the Alps, designed a sickle handle to fit precisely the contours of the hand. Again, his invention appears to have died with him, although today it would undoubtedly win an award at the Design Centre. Once more, those same peasants whom Caesar believed to wear skins invented a very simple beater-in, a comb with short teeth with which to compact the weft in the textiles they were weaving, the same small tool being found universally throughout Britian but nowhere else in Western Europe. These few examples, which could be multiplied many times, should be sufficient to dispel any preconceived ideas about the lack of inventiveness of the prehistoric peasant.

There existed, furthermore, amongst the Barbarians a number of fairly sophisticated pieces of equipment, facets of which we have already noted in our review of the main stream of technological

238 Diagram showing the structure of a roller-bearing from the wheels of a cart from Denmark, about 100 B.C. The device was made of wood and bronze

The craftsmen of barbaric Europe were perfectly capable of designing and constructing quite sophisticated pieces of equipment. This roller-bearing is a good example of the inventiveness of a people whom their more civilized neighbours looked upon as savages.

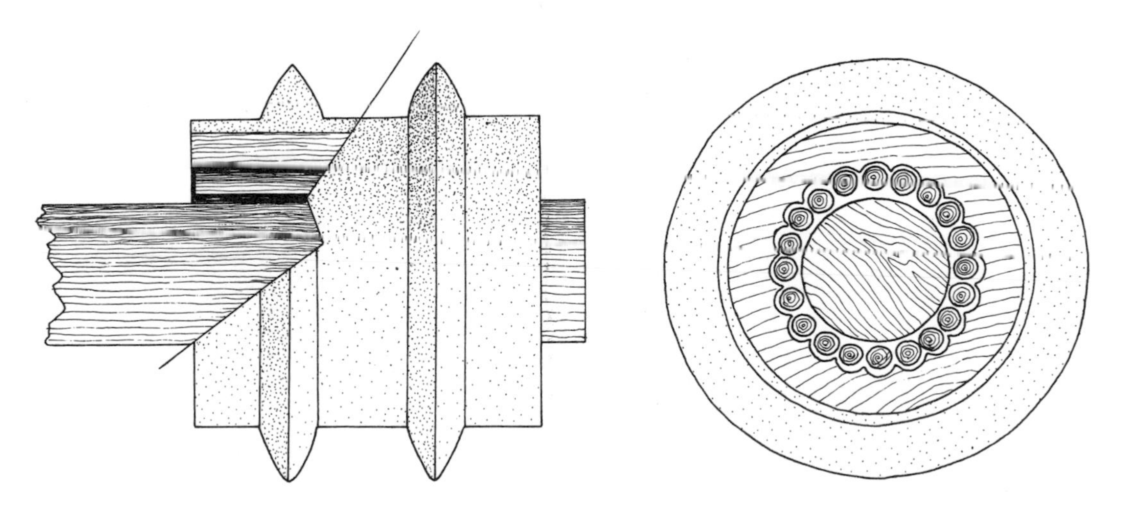

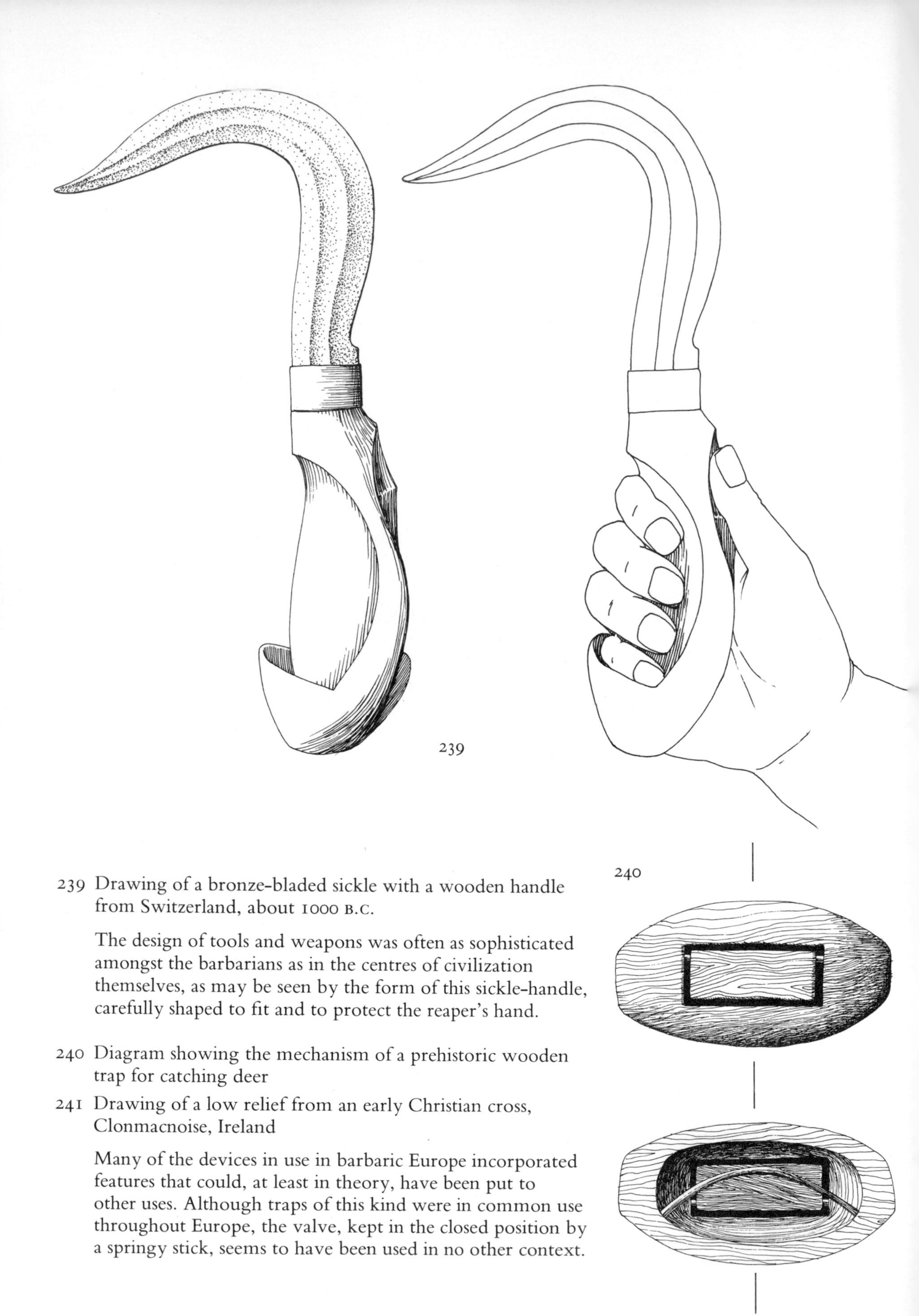

239 Drawing of a bronze-bladed sickle with a wooden handle from Switzerland, about 1000 B.C.

The design of tools and weapons was often as sophisticated amongst the barbarians as in the centres of civilization themselves, as may be seen by the form of this sickle-handle, carefully shaped to fit and to protect the reaper's hand.

240 Diagram showing the mechanism of a prehistoric wooden trap for catching deer

241 Drawing of a low relief from an early Christian cross, Clonmacnoise, Ireland

Many of the devices in use in barbaric Europe incorporated features that could, at least in theory, have been put to other uses. Although traps of this kind were in common use throughout Europe, the valve, kept in the closed position by a springy stick, seems to have been used in no other context.

development in the Near East. Thus throughout a large part of Western Europe there was in widespread use a wooden trap with a spring-loaded valve designed to catch deer. The valve was, in fact, a small hinged door let into a slot cut in a log, and kept in the closed position by a length of springy wood. The whole device was buried in the pathway used by wild deer, and when the beast put its foot on the valve the spring released it sufficiently to allow the foot to sink through the hole in the log where it was held firm by the valve under the tension of the spring. With very little adaptation the makers of these traps could have put this valve to many other purposes had the need arisen, and we cannot even say that the Barbarians lacked the necessary mechanical skill with which to make technological advance. Indeed, one of the most interesting aspects of the study of prehistoric Europe is the way in which various technologies were acquired from the higher civilizations of the Mediterranean and applied by these people to their way of life. What is perhaps yet more interesting is not that many technologies were acquired at all, but the selection of those that were and those that were not adopted. At first sight there seems to be little rhyme or reason in the choice of materials and techniques taken over by the Barbarians from the more civilized world.

In the long period prior to the use of metals the peasants of prehistoric Europe gradually acquired cultivated crops and domesticated animals, as well as many crafts that one would expect to find associated with them, such as the use of millstones, the making of pottery, and

241

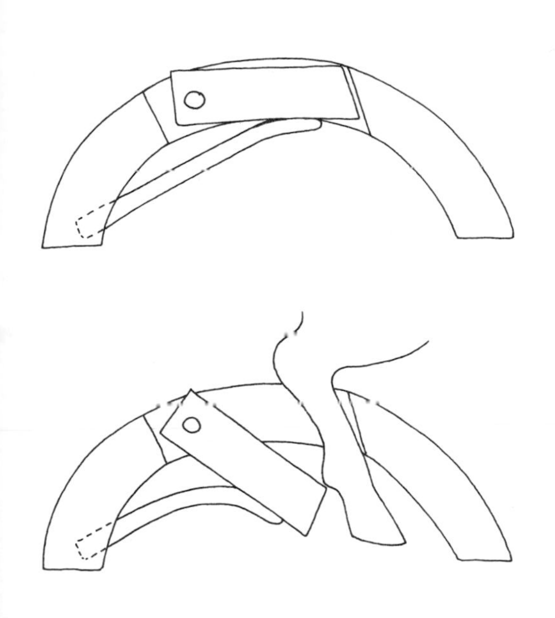

242 Bronze shield from the Thames, late first century B.C.

Metal objects, although never attaining quite the dimensions of those of the civilized world, were quite as sophisticated and as intricate amongst the Barbarians.

fairly complex wood-working based upon the use of stone tools, one aspect of which was the building of elaborate barns and houses. But here the analogy ends. As far as we can judge, many of their villages were of a completely temporary nature and many more were occupied for only a short period of time. The villages, too, were minute by Near Eastern standards and nothing even approaching a small town seems to have come into being.

Later, these same people were to acquire the use of metals; gold and copper in the first instance, and then bronze, with extraction and casting techniques every bit as sophisticated as those to be found in the Near East. But glass-working was quite unknown to them, and their pottery by Near Eastern standards remained for a long period extremely crude. Indeed, during the Bronze Age the most extraordinary situation was to be found, in which clay moulds used for the manufacture of bronze implements were made with the greatest of care, considerable attention being given to the nature of the raw materials used in their manufacture, while at the same time the pottery used by the same people was unspeakable rubbish, thus showing that while they were able to employ ceramic materials intelligently they did not always choose to do so. Accompanying the spread of the use of metals we also find the building of carts and ploughs, but the chariot by contrast was slow in its diffusion throughout the prehistoric period in Europe, and for the larger part its appearance can be equated with that of iron-working. Villages remained small and still of a rather impermanent nature, while the prehistoric Barbarians knew nothing of writing and recording either on clay or on any other material.

The prehistoric Europe into which Rome eventually moved, therefore, was one in which iron technology had spread, in which chariots were still in use as a weapon, at least in Britain, and in which the people used gold coins. The Barbarians had learnt to make pottery on the wheel, and to turn wood on a lathe. In one place they had even made an experiment in building in a totally inappropriate material, mud-brick, a method of construction that they seem to have abandoned equally rapidly. The Gauls and their neighbours were, nevertheless, an illiterate people living in villages. We are probably not far from the truth if we assume that the ability to read and write went hand in fist with city life and that all that was really missing in prehistoric Europe was the appearance of an urban civilization, and we can narrow the argument further to a discussion of why the city never took root in Barbarian Europe.

For a city to come into being demanded a certain degree of social organization and it might be thought possible that in prehistoric Europe such an essential organization did not exist amongst the rather scattered tribes that occupied the territory. But this cannot be totally true. There have survived even to this day a sufficiently large number of elaborately constructed tombs, meeting-places and fortresses, as well as the occasional temple such as Stonehenge, to show that where there was a will the population could be organized in sufficiently large numbers to carry out this type of work. The elaborate banks and ditches around the central enclosure of many fortresses, for example, could not have been carried out without a strong central direction and a very considerable labour force.

Perhaps we will understand the non-development of the city in prehistoric Europe better if we look briefly once again at the development of the city in the Near East. By many writers this development has been described as essentially the growth of a complex of dwellings around a central temple, shrine or place of worship; and this indeed is often the superficial appearance given by such cities upon excavation. But in fact most cities were primarily the centres of particular technologies. Throughout the whole period covered by this book most cities were renowned simply for the presence within their walls of one or two main technologies such as the making of pottery or glassware, or the manufacture of jewellery, and there is very good reason to believe that more often than not it was these industries, and not the presence of the temple, that brought the city into being. Of course, where a technology developed there had to be a means for the import of its raw materials and the export of its finished products so that the city became equally a centre of trade and a market-place for the buying and selling of agricultural products and livestock to provide food for its citizens. This argument must, therefore, lead us to the point of view that in some degree the main distinction between technological development in the Near East and in prehistoric Europe was that the Barbarians failed to centralize their industries.

To a limited extent this point of view is borne out by the nature of those technologies absorbed by the Barbarians and of those which they apparently rejected. For example, it was not necessary to construct an elaborate foundry for the production of bronze weapons and tools. The

243 A collection of bronze scrap-metal from Saxony, about 700 B.C.

The barbarian bronze-founder was essentially an itinerant, setting up a temporary work-shop as and where required. Dumps of scrap-metal, previously collected from customers and hidden for future use, have often been discovered, normally unrelated to buildings or settlements of this period.

bronzesmith himself might have been completely mobile, shifting from one community to another, on each occasion setting up a temporary furnace to produce the objects required. The fact that dumps of broken bronze objects, clearly intended to be resmelted, have frequently been found unrelated to any building shows that in prehistoric Europe the bronze industry was operated largely in this way. On the other hand, glass working did require a more permanent set-up, and although just possible, it would have been extremely difficult for a glass-worker to move all his equipment from one community to another regularly throughout the seasons of the year. Equally, as we have seen, the manufacture of crude pottery could have been carried out with few or no large tools at all, while the production of sophisticated pottery demanded a lot of permanent equipment such as wheels on which to throw the pottery, and kilns in which to fire it. In the absence of urban centres from which its products could be disseminated, we would, therefore, not expect to find wheel-thrown pottery.

Even in those areas from which raw materials were acquired in bulk, as for example around copper- and iron-workings, salt mines or tin mines, no cities seem to have developed, and it would appear that in Barbarian Europe raw materials were acquired and distributed in the same *ad hoc* way as that in which the craftsmen moved from one community to another. Yet the volume of trade was often sufficiently large to warrant urban development and at times it would seem almost as though there were amongst the prehistoric people of Europe a deliberate attitude of thought that ran counter to building cities. In Britain, for example, the Roman policy of creating towns was a signal failure: the local population flatly refused to live in them, and often cities that were planned were never completed for lack of occupants.

Whatever the cause, and it may in the end have been merely that the population in Barbarian Europe was insufficient to support cities, their failure to produce urban communities can be seen as a primary cause for their failure to produce or to adopt new techniques. Our past review of the history of technology in the ancient world has shown the enormous reliance of one technology upon another, and in scattered communities, where there could have been little or no exchange of ideas, new technologies are likely to have fared poorly and inventions to have died stillborn.

The Barbarians in the East: The Indus Valley

So far we have failed to record adequately the fact that the city settlements in the Valleys of Mesopotamia and of the Nile had their parallels in the Indus Valley. Indeed the archaeological record shows that farming communities, similar in general pattern to those that we have seen in the Near East, had taken root in this part of North-west India well before the year 3000 B.C. Later were to develop a number of cities of which Harappa and Mohenjo-daro were the largest and are probably the best known. In many respects these cities and the technologies practised within them were a reflection of what we have already seen in the Near East. Building in mud and later in fired brick; the development of a system of irrigation; the smelting of copper and then bronze; the use of Egyptian faience and of stamped seals all appeared in much the same order as they have already been seen to have appeared in Egypt and in Mesopotamia, while in many other respects the civilization of the Indus Valley followed very closely that of Mesopotamia. The cubit and the foot, for example, of identical magnitude to those used in Mesopotamia, were the current measurements of length. One might even be excused for supposing that the people of the Indus Valley were not particularly inventive and borrowed heavily from their distant neighbours in the West, and that the Indus Valley was a cul-de-sac which accepted ideas but produced no new technological innovations itself. This point of view, however, needs very considerable modification for they were not totally bereft of their own technologies.

The people of the Indus Valley certainly cultivated new crops and domesticated other animals than those found in the Near East. Shortly after 3000 B.C. cotton was being cultivated as a fibre plant for making textiles, while the humped cattle of India, for which it must be said that no wild prototype is immediately apparent, were in domestic use at an equally early date. The fact that their origin is obscure suggests that stock-breeding had been practised in the Indus Valley for a long time before this. While the people of the Indus Valley bred all the farm animals common at this time in Mesopotamia, they had also domesticated the buffalo and tamed the elephant.

If the use of stamped seals and the basis of linear measurement were borrowings from Mesopotamia, the same cannot be said of their script, although of course it is unlikely that the concept of writing was an independent invention of the people of the Indus Valley. The script itself is totally different from that of Mesopotamia or of Egypt and it remains largely undeciphered. Although clearly based on some form of pictogram, the symbols as we know them are a long way removed from the type of pictogram that illustrates some common object. It is just possible of course that earlier forms of this script are yet to be discovered, but at the moment it looks very much as though this form of writing was evolved fairly rapidly to meet an immediate demand. In these circumstances we must suspect that the concept of writing was lifted wholesale from Mesopotamia, and that in the Indus there were no early evolutionary steps.

Whatever the connexions, however, between Mesopotamia and the Indus Valley they can never have been very strong. Admittedly seals and other small objects of Indian origin have turned up in Mesopotamian cities, and vice versa, but there are nevertheless extraordinary lacunae in the development of many technologies in the Indus Valley that could never have existed had there been strong connexions

244 Two seals from Mohenjo-daro, about 2000 B.C.

In the third millennium B.C. a number of cities developed in the Indus Valley with technologies very similar to those of contemporary Mesopotamia. In India, however, the local fauna was exploited, as for example the elephant and humped cattle.

The form of writing seen on these seals (still largely undeciphered) was clearly based on earlier pictograms. Its relationship with that of Mesopotamia, if any, remains obscure.

between the two areas. Thus, for example, we find in the field of metal technology first the use of copper and then at a later date the manufacture of alloys, a development strictly comparable with that in Mesopotamia. This being the case, one might expect to see the same sequence of events in the actual design of moulds into which to cast the metal, but curiously enough the open mould continued in use over a very long period. It was not until almost the dramatic end of this civilization that piece-moulds were introduced and even then they were undoubtedly brought in by intruders from the north.

More than one archaeologist writing about these early cities in the Indus Valley has commented upon the dreariness, the repetitiveness and the dead-level nature of the various objects produced by the Indus craftsmen. Looking at the way in which the cities were planned, with their rows of mean huts for the workers, grouped close to furnaces for metal production or to pottery kilns, one has the uncomfortable feeling that here were city states in which production was ruthlessly organized but in which the techniques employed were probably not very efficient. One gets, in fact, the feeling that here again the dead hand of the civil servant was in operation such as was surmised during the declining years of Rome.

In discussing Barbarian Europe, it was suggested that the failure of the primitive peoples of prehistoric Europe to produce new technologies was due to a lack of cities. If the Indus Valley civilization is to tell us anything, it illustrates perhaps the fact that cities alone are insufficient if technologies are to be thriving and progressive. For the great cities of Mohenjo-daro and Harappa were as large and as industrious as any known from the ancient world at this date; and while we may blame the way in which their industries were organized, in all probability the root of the trouble lay in the difficult communications between the Indus Valley and the rest of the civilized world. As it is, it would seem that these two great cities and their dependent towns and villages became technologically introspective. They had achieved a level of technological development that was satisfactory from their point of view, and there appears to have been inadequate contact with the rest of the civilized world to act as a stimulus to further development.

Many of the achievements of the Indus civilization were abruptly destroyed somewhere shortly after the year 2000 B.C. when Northern India was overrun by groups of people from the north akin to those who, as we saw in an earlier chapter, overran Greece and Anatolia. These tribesman were equipped with superior weapons including the chariot: the cities of the Indus were destroyed as, too, was their administration. It was left to the invaders to choose which of the existing technologies that they found in the Indus Valley they would use, but so far as the history of technology in the ancient world is concerned, the contribution from Northern India was to be negligible. It was not until after more than a thousand years that one of the Indus Valley's independent developments, cultivated cotton, was imported into first Mesopotamia and then Egypt.

The Barbarians in the East: China

Our understanding of the development of civilization in early China is today extremely scrappy, due almost entirely to a lack of archaeological excavation designed to throw light upon the period of its origins. Although much has been done in the last few decades to remedy this situation, it is still almost impossible to write with conviction about the earlier periods covered by this history. Thus, for example, almost nothing is known of how agriculture and stock-rearing came to be practised in early China. But within the few centuries immediately before the year 2000 B.C. there were clearly a number of agricultural settlements in the main river valleys, and while we know very little indeed about the economy of these societies two aspects of their technology, or what remains to us of their technology, are of considerable interest. On the one hand we find that these people were making pottery by the same techniques that had been used in Western Asia, but the style of the wares they were producing was particularly their own as, too, were the forms of decoration. The pottery was made on a turn-table as it had been in Western Asia; it was decorated with pigmented clays painted on to the surface; and it appears to have been fired in some sort of primitive kiln. There is, however, nothing in the Chinese archaeological record to suggest that the art of making pottery was a local development, and we may suppose, for want of evidence, that the technique was initially acquired from the Western Asiatic societies that we have already discussed.

A second aspect of Chinese technology at this early date is of particular interest not so much because of the techniques being employed as because of the raw material used, for amongst the objects from these early agricultural settlements are a number of plaques made of jade.

245 Reconstruction of the kind of Chinese mould used for casting a bronze vessel of about 1500 B.C., based on existing fragments of moulds

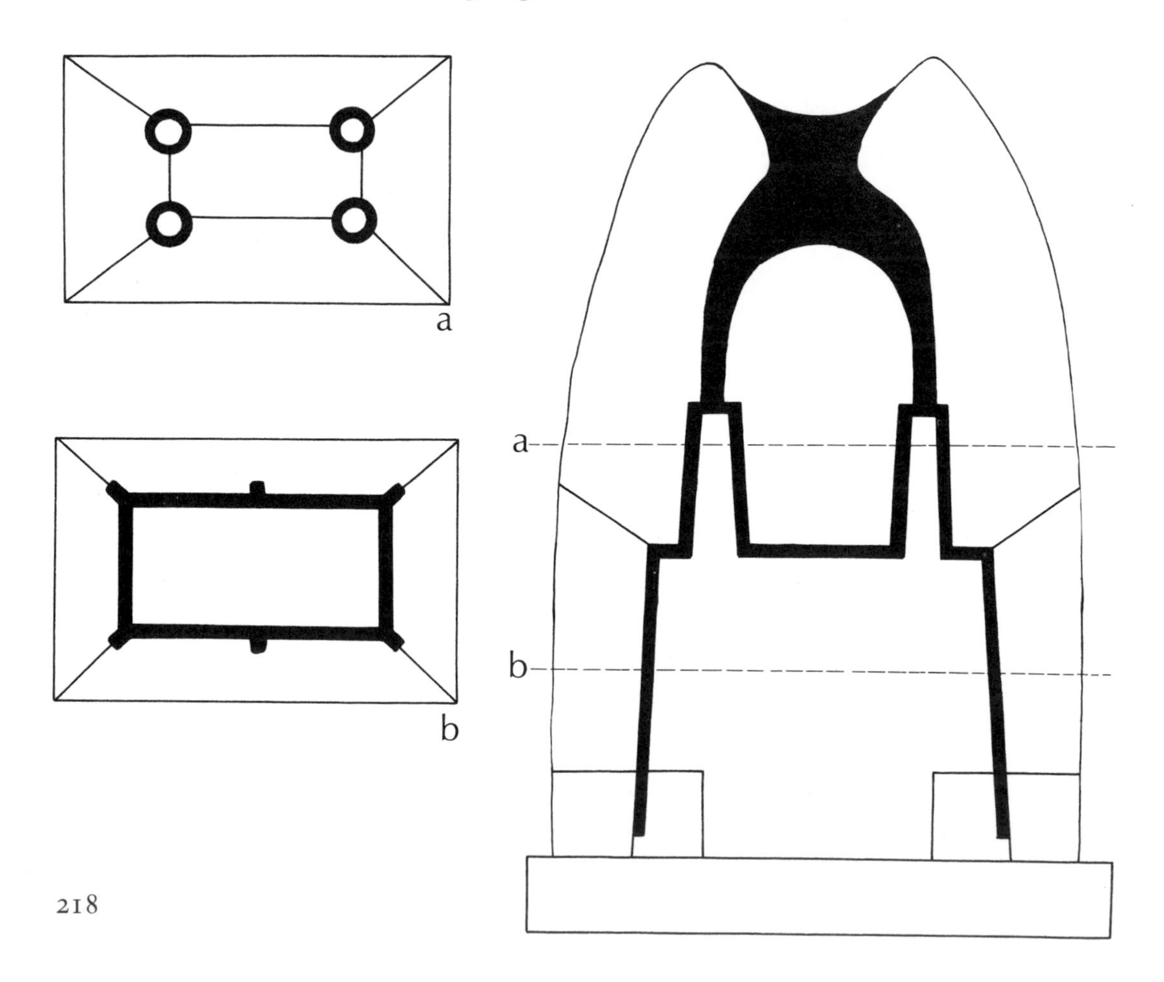

This is a hard rock and difficult to work without metal tools: it is also a mineral of limited origin. It does not occur as a deposit within China proper, and although it has been argued that the stone used was in the form of geological erratics, the argument carries little conviction. The most likely source, in fact, would seem to have been many hundreds of miles from China, in the mountains of Sinkiang. The exploitation of a rock so far from its source by a simple agricultural community is, of course, remarkable; but the fact that this was so serves as a clue to the entire problem of the early technology of China. The trade route, stretching across Asia from China to the cities of the Near East, three thousand years later to be known as the 'Silk Road', must to some limited degree already have been in use, and we are left speculating by whom and for what reward jade was carried this vast distance into China at so early a period.

What has been said about the appearance of the technique of potting in China is equally true of the practice of bronze-founding, for about the year 1500 B.C. bronze appeared, so as to speak, out of the blue. There have been found no early attempts at metallurgy, no unalloyed copper tools, none of the simple forms of stone mould that we have learnt to expect in the West. Instead one finds that the Chinese were suddenly producing bronze vessels of very considerable complexity using elaborate methods of casting. We must suppose, therefore, that the technique of casting, the knowledge of how to win the metals from their ores, and to form alloys, were introduced from without, presumably by the same route as that which brought jade. Even so, from the very beginning the Chinese approach to bronze-working was totally different from that of Western Asia.

In Western Asia early metallurgists treated their metals as though they were some form of superior stone, and although they were cast into moulds to produce the approximate shape of the tool or ornament required, most of the finishing was done later by hammering and polishing, and in very few cases until the later periods does one find moulds designed so that little or nothing needed to be done to the surface of the bronze after it had been cast. By contrast the Chinese attitude towards the metal was that of the potter: it was treated like some superior ceramic. The Chinese spent most of their effort in creating moulds that were so precise in every detail that, once cast, the bronze object needed little or no fettling. In order to achieve these results, the Chinese metallurgists found it necessary to raise the lead content of their bronzes, and it is not at all uncommon to find the lead content of the alloy as high as fifteen per cent, while on occasions this proportion rose to thirty per cent.

The precise method by which early Chinese moulds were made has been the subject of heated debate, some claiming that the whole production was achieved by the use of a great number of extremely carefully fitted pieces into which had been impressed all the details of the objects to be made. Such piece-moulds might contain thirty or more sections, all of which would have had to be keyed together before the casting was made. Indeed, a number of fragments of such moulds have been found, which alone lends support to the view that this was the technique employed. On the other hand, some scholars, after examining surviving bronzes, have come to the conclusion that they could have been made only by the lost-wax method. On the face of things it would seem probable that both methods of production were being used concurrently, although the piece-mould seems to have been the more common of the two, and it may well be that those who adhere to the view that many of the bronzes were cast by the lost-wax method have failed to take into account the enormous strides made by the Chinese in the field of ceramics, and have underestimated the ability of the Chinese to make such intricate moulds at this early period.

At the same time that bronze was first used in China, very considerable advances had been made in the field of pottery production, and the local craftsmen were employing one of those rare mineral deposits, china clay or kaolin, in the manufacture of pure white wares. The method of decoration applied to this pottery is curious, for far from being cast, as was the decoration of the bronze vessels, it was carved into the surface after the wares had been allowed to dry, but before firing. This is the more curious since the motifs used on both the bronze

246 Chinese bronze and pottery vessels of about 1500 B.C.

247 Crucible used for casting bronze, from China, about 1500 B.C.

In China no early stages of the development of metallurgy are known: the appearance of bronze seems to have been quite abrupt. The techniques of working the metal, however, differed from those in contemporary Western Asia, as, too, did the purpose to which it was put. In China the metal was cast into very complex and accurately made moulds so that a minimal amount of hammer-work was required in its finishing. A further contrast may be seen in the type of crucible used which was made with a double wall to insulate the wooden tongs used to grasp it. Curiously, the Chinese of this period do not appear to have made pottery in moulds, despite the fact that bronze and clay vessels were often remarkably similar in form.

246

247

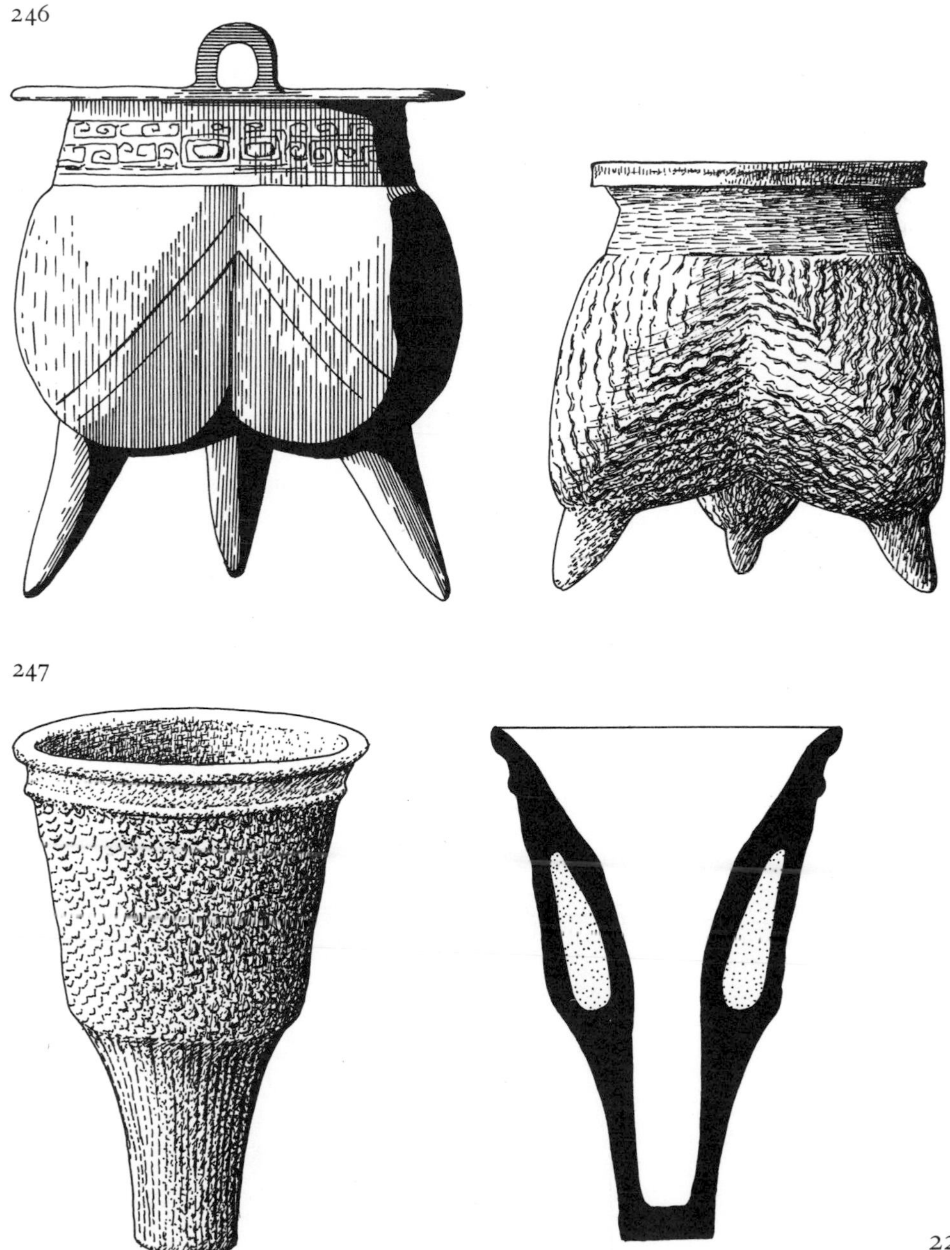

and pottery vessels, although seldom identical, are often of a similar nature. Indeed the whole style of decoration is reminiscent of flat carving such as might have been carried out on the surface of a vessel made of bamboo, wood or even bone. Apart from the early exploitation of china clay, felspar, previously ground to a fine powder, was being used to form a blotchy and uneven glaze. Later it was to become the basis of glazes applied to Chinese stoneware. Although we know nearly nothing of the kilns of these early potters, it is quite evident that they must have been superior in many ways to those being used in Western Asia, capable of achieving rather higher temperatures and of greater control of the atmosphere within them. For some unexplained reason neither the white wares nor the felspathic glazes remained in vogue for long, and by the year 1000 B.C. both had become uncommon and were shortly to disappear, not apparently to be made again for nearly another millennium.

Bronze, however, was not the only borrowing from the West, for during this period the Chinese adopted two important weapons, the chariot and the composite bow. The chariots, unlike those we have seen in Western Asia, retained their axles in the middle of the platform, but the wheels were often fitted with as many as eighteen spokes. They

248 Reconstruction of a Chinese chariot of about 1000 B.C., based upon excavated examples

The Chinese appear to have acquired the chariot from their nomadic neighbours at much the same time that they borrowed the composite bow, although both seem to have been adopted later in China than in Western Asia. The earliest Chinese chariots were fitted with multi-spoked wheels, but the yoke and other fittings differed little from those in the West.

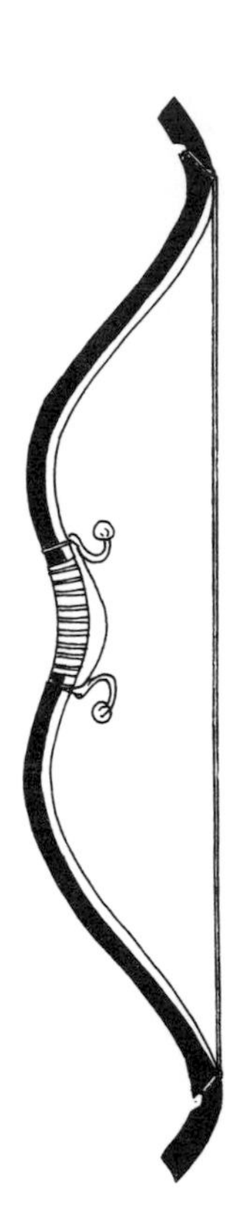

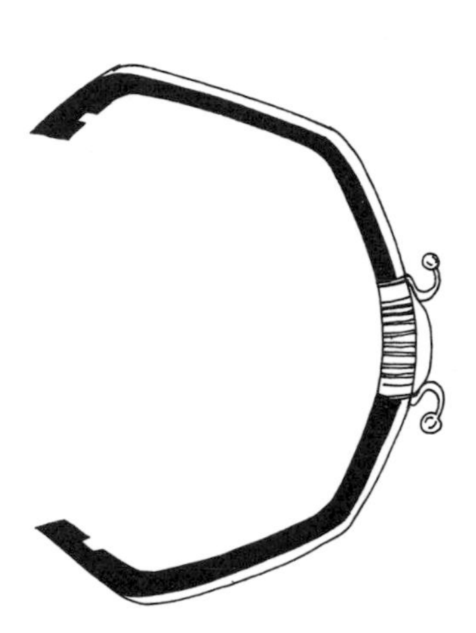

249 Reconstruction of the Chinese composite bow of about 1000 B.C. based on surviving fragments

The composite bow, made of lamellae of wood, or of wood and horn, seems to have been developed by the Asiatic nomads. A short weapon, but nevertheless with a considerable range, the composite bow was ideally suitable for use from a chariot or from horseback where a longer bow would have been too cumbersome.

were, nevertheless, pulled, as in the West, by horses in paired draught using the yoke in the form of an inverted letter 'Y'. The composite bow, made of separate lamellae of bone, wood and horn, although comparatively short and therefore easily handled from a chariot, provided as much power as the longer wooden bow of the typical infantryman. It, too, appeared in Western Asia rather earlier than in China.

If bronze, the chariot and the composite bow all appear to have been importations from the West, the same cannot be said of early forms of writing amongst the Chinese. Certainly the first characters of Chinese writing were pictograms, as they were in the West, but they were not used for the reason of keeping records, nor were they written on clay tablets or on papyrus. They were, in fact, made on bones the purpose of which was divination. It would seem that no one of any standing would have dreamt of starting a new enterprise without first consulting a soothsayer. His intention was inscribed upon the bone, which was then prodded with the red-hot point of a bronze tool, and depending on the way in which the bone cracked, it was decided whether the omens were good or bad. Originally there were some five thousand symbols, of which we can now only decipher about a third. This is largely due to a complete reform of the system of writing in China in the second century B.C., but those pictograms that we can understand formed the basis of the later script.

Although there seem to be many points of similarity between the development of technology in China and in Western Asia during this period, there was a very marked difference in the organization of workshops and industry generally. Broadly speaking, in China the whole of industrial output was in the hands of the rulers, and while in the West bronze was used for tools as well as for weapons, in China almost the entire output of the early workshops went into the manufacture of ritual bronze vessels and, of course, weapons. The farmer, the woodworker and indeed most other craftsmen still had to rely upon stone tools, and it was not until the appearance of iron that the ordinary artisan was able to acquire metal tools.

When eventually the Chinese took to using iron, the entire process of its preparation was totally different from that in Western Asia, for the earliest iron objects that we know of, from the fourth century B.C., were made not of wrought metal but of cast iron. It will be remembered that in the West iron ore was first reduced to a bloom, and that this was then hammered up into wrought iron on an anvil. In China, by contrast, the ore was reduced directly to the molten metal which was poured into prepared moulds, thus directly continuing the long tradition of bronze-casting. This was possible only because of a number of local circumstances. First of all, the majority of the iron ores available in

250 An early pictogram of a plough from China, about 1500 B.C., and a drawing made from a low relief of about 200 B.C.

Early forms of plough in China differed in design, but not in function, from those in use in Western Asia.

The earliest forms of writing known in China were, as in the West, pictograms, but the symbols were engraved upon bones which were used to foretell the future. The later Chinese brush-and-ink symbols were derived from the pictograms.

251 Chinese tools of cast iron with reconstructed wooden handles and blades, about 400 B.C.

Before the use of iron, the Chinese craftsmen had to rely entirely on tools of wood, stone and bone, for bronze was reserved exclusively for ritual vessels and for weapons. The earliest iron used was the cast metal which was too brittle for making weapons, but could be used for tools. Normally cast-iron blades were employed as a cladding to existing forms of wooden implements.

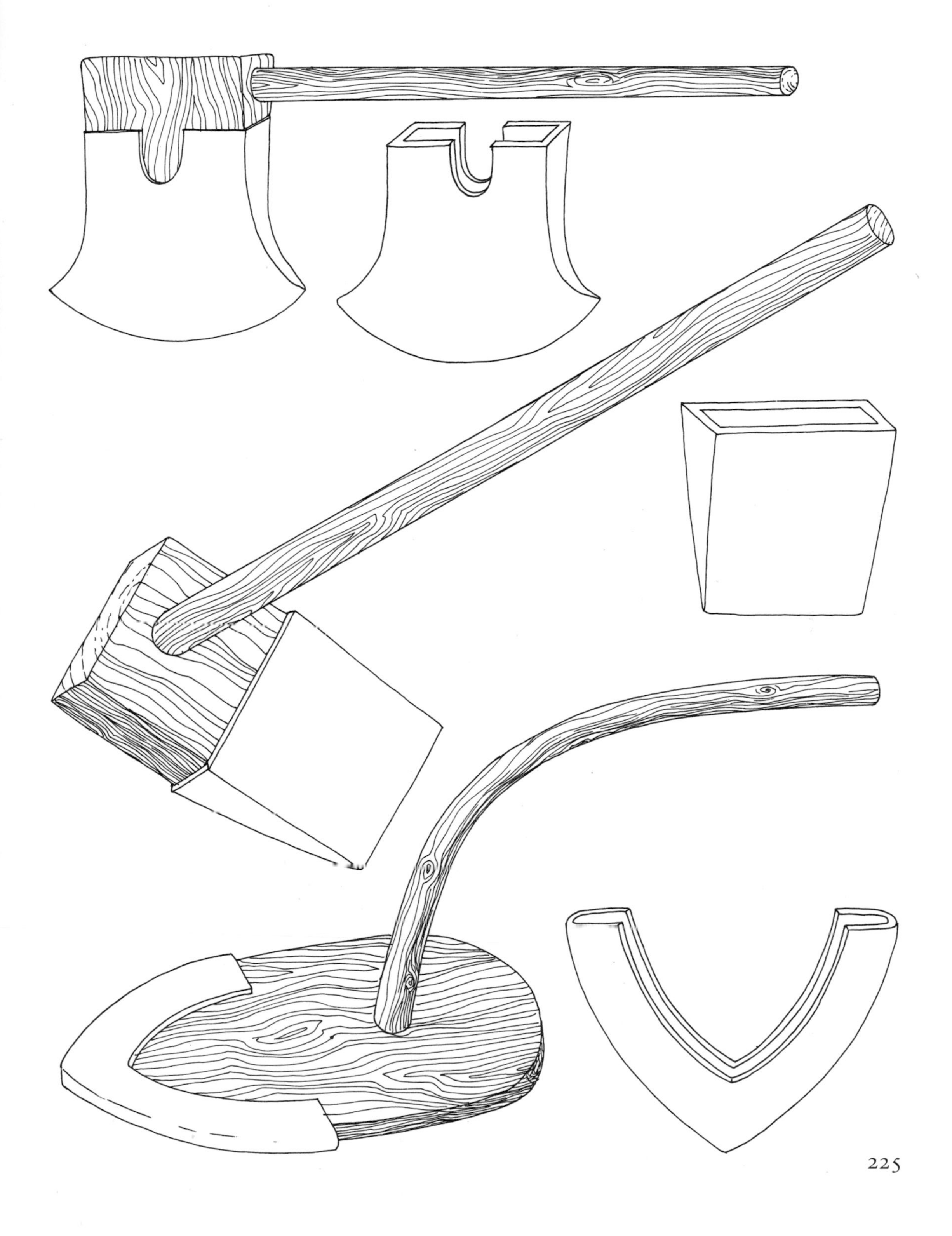

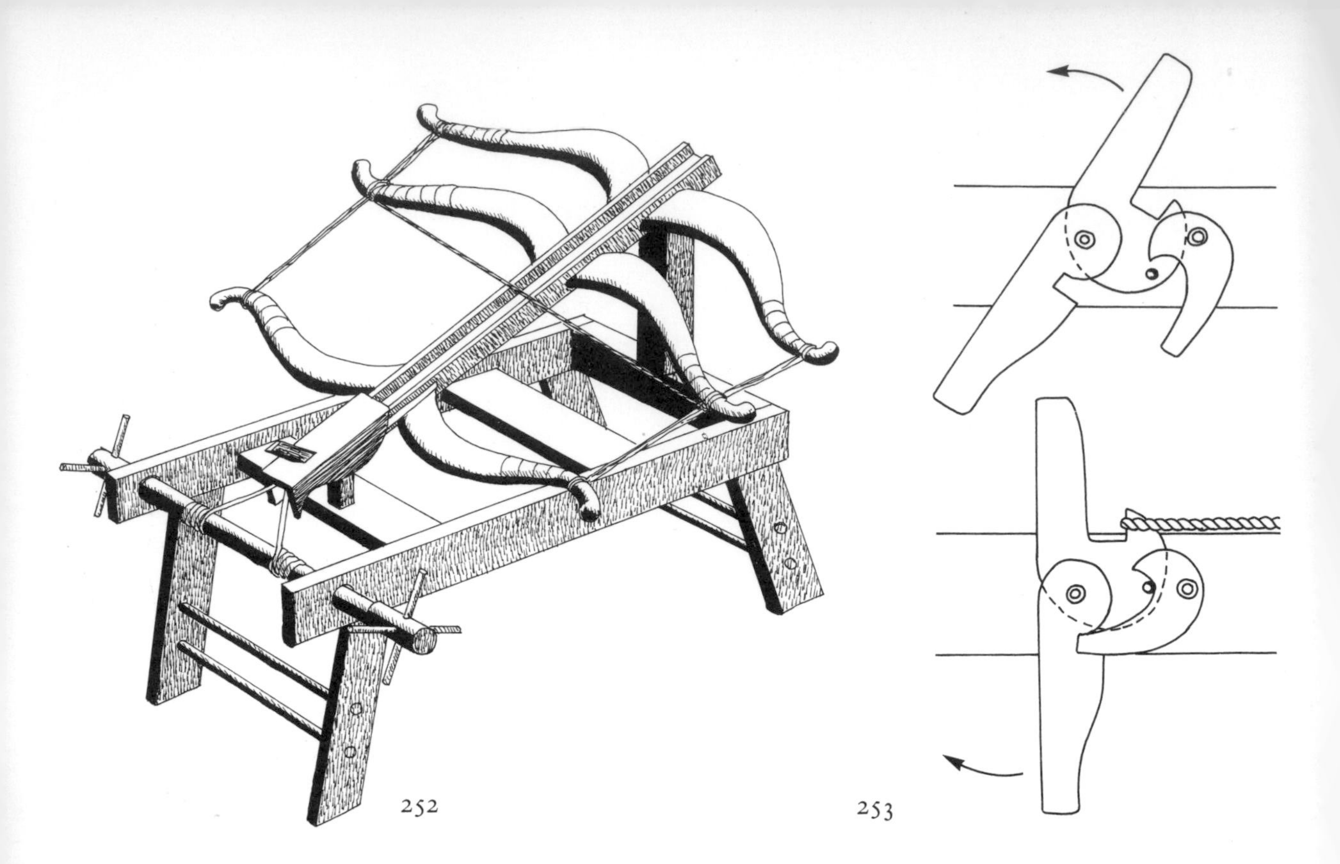

252 253

China had a high phosphorus content which greatly reduced the melting-point of the metal; secondly, there was readily available good refractory clay from which to build furnaces able to withstand the high temperatures involved; and thirdly, the Chinese had developed an ingenious form of piston bellows that allowed them to produce a fairly steady draught through the furnace. The cast iron so produced, however, was far from being universally serviceable, for it was far too brittle for the manufacture of arms and weapons, and it was therefore used largely for vessels and for agricultural tools. It was two centuries before the Chinese learnt how to treat their cast iron to produce a malleable form of metal that could be used for weapons.

During this period the Chinese were under constant pressure from the Nomads on their frontiers, and the incessant incursions of hordes of mounted men armed with the composite bow caused enormous havoc. In response to this threat the Chinese developed the cross-bow which, because it could be held steady and had a greater fire-power than the weapons of their adversaries, more than evened up the odds against the Chinese. Undoubtedly the most ingenious part of the cross-bow was the cocking and trigger device without which it could never have been an efficient weapon, and it is perhaps of interest to note that this same mechanism is to be seen almost unaltered in hammer-operated guns of very much later periods, and indeed as a release mechanism in many early clocks.

252 Reconstruction of an early Chinese cross-bow, based on a drawing of the fourth century A.D.

253 Diagram showing the cocking and release mechanism of a Chinese cross-bow, based on a surviving example

The cross-bow was developed in China in the third century B.C. as a measure of defence against the ever-present hordes of highly mobile nomads. This weapon, with a far greater range than that of the composite bow, originally had a single bow-spring, later to be made in the multiple form illustrated here. The efficiency of the cross-bow depended entirely upon the cocking and trigger mechanism without which loading and firing would have been fatally prolonged. The cocking device in only slightly modified form is to be seen in many fire-arms today – including children's cap-pistols!

254 Diagram showing the probable working of Chinese lantern-bellows of the second century B.C., based on a low relief

255 Diagram showing the working of Chinese double-action piston-bellows of the second century B.C., reconstructed from documentary sources

Efficient iron-working was only possible with improved bellows with which to maintain a steady draught through the furnace. Early forms of bellows from China seem to have been similar to a modern folding lantern, the draught being created by compacting the bellows. Later, with the introduction of the double-action piston-bellows, an even more regular draught could be supplied, so eventually permitting the manufacture of steel from cast iron.

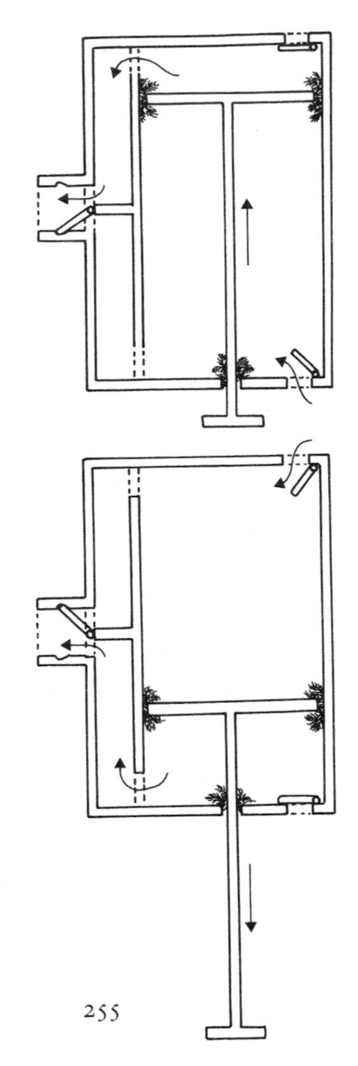
255

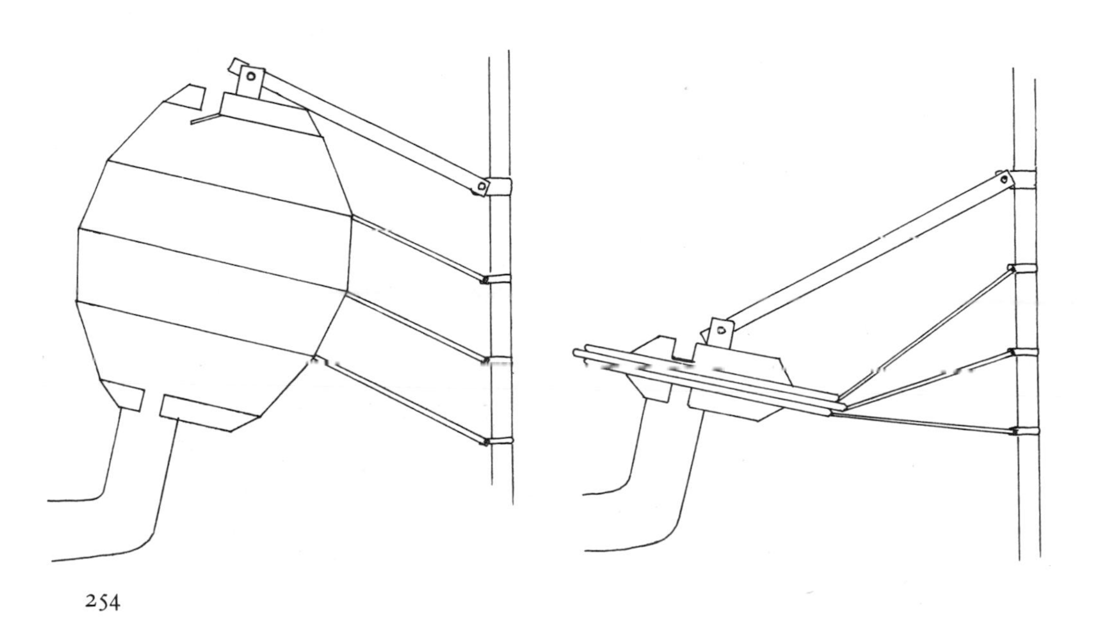
254

256 Diagram showing two types of mill in use in China before the beginning of our era, both based on documentary sources

Early Chinese mills were often similar to those found in Western Asia during the same period, and the same devices may still be seen in use in many parts of the world today. The foot-operated hammer-mill, seen here on the left, was possibly what inspired the iron-forging hammer, ultimately driven by a cam on the shaft of a water-wheel. This allowed the making of larger forgings that would have been possible with hand-held hammers.

At much the same time in China appeared true draught harness for horses – shafts, traces and padded collars – all of which allowed a more efficient use of these beasts of burden, eliminated the need for paired draught and, because there was no further risk of half throttling the horse, gave chariots and carts a wider field of operation. It is, however, far from clear whether this development was truly Chinese: it seems more probable on the face of it that here, again, was a direct borrowing from the Nomads.

In the second century B.C. new impetus was given to the iron industry by the introduction of a double-action piston bellows permitting a very regular draught through the furnace. It also allowed 'fining' of cast iron, a process in which the molten metal was submitted to a steady flow of air, so removing much of the carbon which had made it brittle, to provide a form of malleable iron that could be used for the manufacture of weapons. Much of the hard work was taken out of the actual process of forging by the use of the trip-hammer. This had probably already existed as a form of pounder for crushing grain, being little more than a foot-operated pivoted lever with a heavy hammer mounted at the far end.

The same period saw the introduction of a number of related pieces of equipment that were to prove very important in the future development of Chinese technology. The lap-wheel, operated in a very similar manner to the lathe that was being used in the West, now became the general tool for the shaping of jade. A free-standing wheel was used for the reeling of silk, while a large wheel fitted with edge fans was used for winnowing grain. Two centuries later the latter device was to be adapted to result in an even more efficient iron industry. With the introduction of the water-wheel, the winnowing fan was converted into a blow engine that gave an even more regular draught through the furnace and fining ovens, while the trip-hammer, instead of being

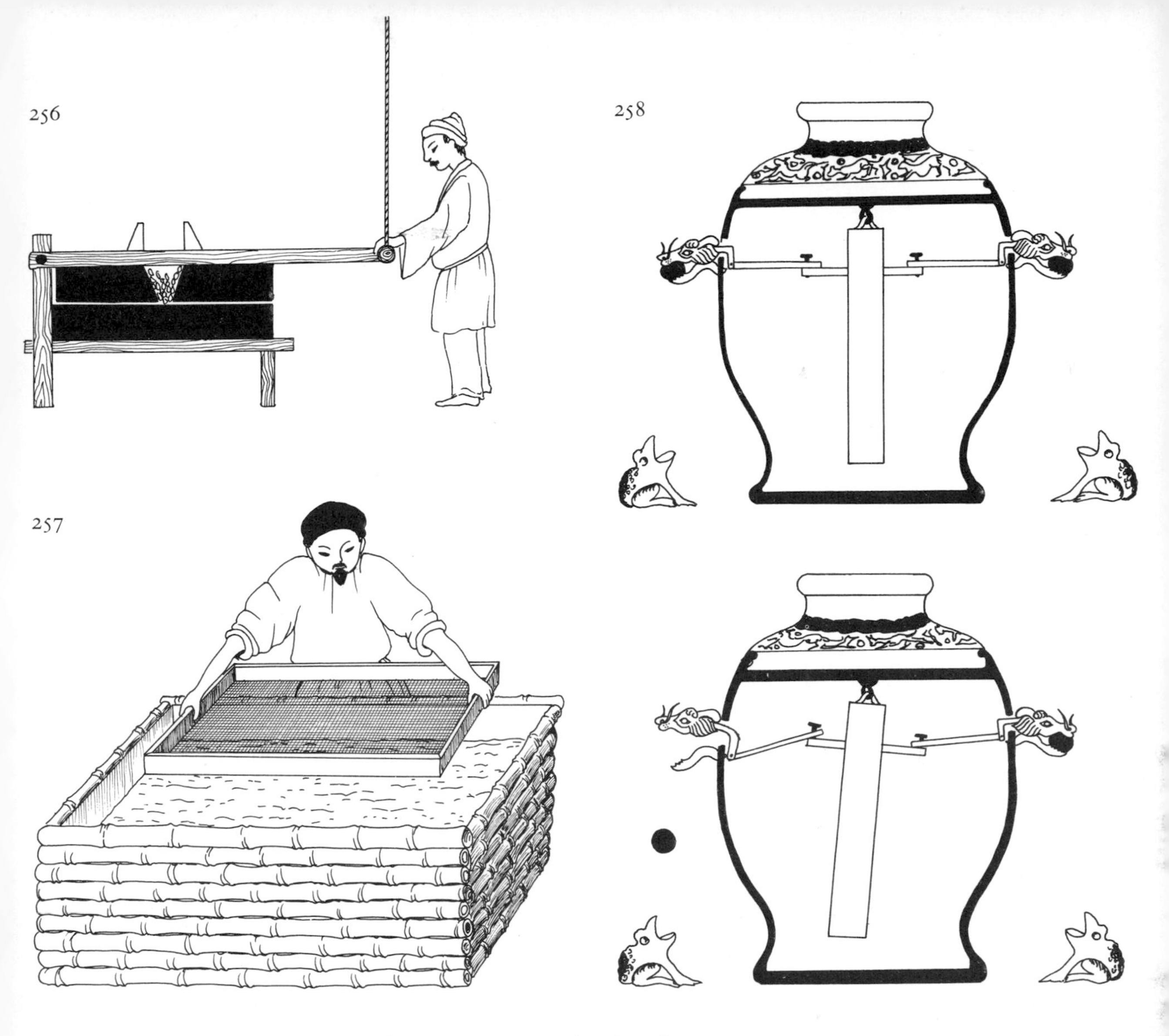

257 Making paper in China, a reconstruction based on a drawing of about A.D. 500

Paper was first made in China in about A.D. 100. From a prepared mash of plant fibres a small quantity was shaken into an even layer on the mesh of a sieve. After draining the paper was removed and stuck against a wall to dry. For centuries before this the Asiatic nomads had been making felt from animal fibres, and it is debatable to what extent paper-making was inspired by the process of felt manufacture.

258 Sectional drawing of a Chinese seismograph of bronze, probably about A.D. 600

This device for detecting earth tremors may be seen as typical of Chinese mechanical inventiveness of this period. Seen in the round, the vessel carried on its perimeter a dozen figures of frogs each with a metal ball held in its hinged jaw. Earth movements would have disturbed the central pendulum sufficiently to release a metal ball, thus showing the direction of the tremor.

worked by foot was now worked by cams on the shaft of the water-wheel. By the fifth century A.D. the Chinese were producing both cast iron and wrought iron and, by co-fusion of the two metals, were able to provide good-quality steel. This industry was, hence, far in advance of that in the West.

Space does not allow us to elaborate upon the many other inventions of this epoch in China, but the first five centuries A.D. also saw the reintroduction of stoneware, now in a far more refined form, and the invention of paper. The latter was made from a mash of plant fibres which was spread evenly on the bottom of a fine sieve, allowed to drain, and the thin sheet so produced then stuck against a wall in the full sun to dry. At first sight this would appear to be a totally Chinese development, but the fact that the Nomads had long since been using a felt made of animal fibres should not be overlooked, for in essence paper is merely a felt made of vegetable matter rather than of hairs.

Throughout this all too brief survey of the early development of Chinese technology we have frequently referred to the Nomads, as we did when speaking of the main stream of technological development in Western Asia, and it is now to these people that we must turn.

The Nomads

At the very beginning of this book the reader was warned that the history of technology in these early periods would, through lack of knowledge, have to be incomplete, and we now come to one of the largest gaps in our history, for what can honestly be said about the people of the steppes is remarkably little. The reason for this lack of information is simple. They were a people largely without cities, and most of our knowledge of them comes from the excavation of graves. Many of these were marked by a tumulus raised over the burial-place, and these have of course attracted the attention of excavators. But many were not marked in any way and their discovery has been purely a matter of chance. Furthermore, such cities as existed in all probability occupied precisely the same sites as those of modern cities. There is, for example, no knowing how far back such a town as Samarkand could be traced in history were it possible to excavate under the present buildings. As things stand, however, we must make do with what little material is available and be forewarned that most of what follows must be conjectural.

We should not, however, allow ourselves to be confused by the word

Nomad. Today we often use it to describe someone who is foot-loose or we think of gypsies or other relic populations who have maintained a mobile way of life and no longer truly fit into modern society. The word is tainted and has the implication of a people who are backward or even dishonest. In the period of which we are speaking, however, there is no good reason to believe that the Nomad was in most respects less advanced technologically than his sedentary neighbours, while the fact that he followed his herds and has left behind no obvious traces of towns or cities should not allow us to think of him as someone substandard. From the very moment that wheeled vehicles came into being the covered wagon and the tent became the Nomad's house, and the few examples of such vehicles that have come to light as a result of excavation demonstrate that they were no better and no worse equipped than the mud-brick dwellings of the average peasant in Egypt or Mesopotamia. They were if anything probably rather more hygienic. Essentially the Nomad encampment was itself a city with many of the appurtenances – both social and material – to be found in the more settled communities of this epoch.

Nevertheless the way of life imposed upon the Nomads by their need to be on the move in search of fresh pastures for their cattle meant that in certain fields technological advance was bound to be inhibited. Evidently we would not expect, nor do we find, complex building techniques amongst these people, and by the same token, since they were not forest-dwellers they were not involved in the creation of large wooden structures either in the form of buildings or of mechanical devices. Equally we would not expect the Nomads to be responsible for the production either of sophisticated ceramics or of glassware. Nevertheless when we attempt to assess the role of the Nomads in the development of other technologies such as metal-working or the evolution of the wheeled vehicle the situation is totally different, and our problem is chiefly to assess whether the Nomads were responsible for certain innovations or merely the borrowers and transmitters of ideas.

As we have already seen, the carts and wagons at first known in the Near East were already fairly advanced. The earliest wagons which we know of on the Asiatic steppe appear to have been of a comparable date, and so far as our knowledge of these things goes it would seem more than likely that vehicles with solid wooden wheels and drawn by oxen in paired draught were in fact an invention of the Nomads themselves, an invention later borrowed by the people of Mesopotamia. At a later date the taming and breeding of horses was certainly

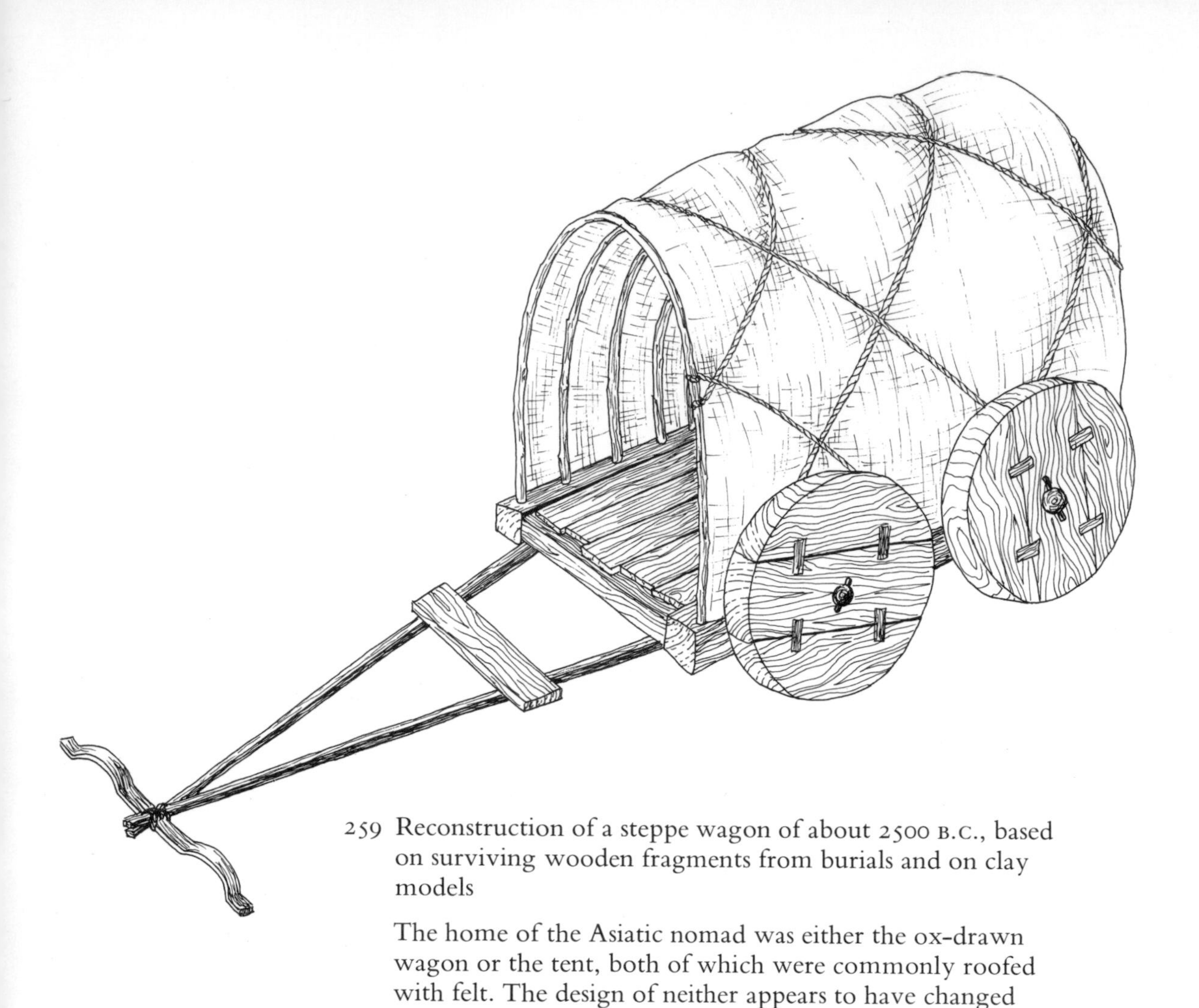

259 Reconstruction of a steppe wagon of about 2500 B.C., based on surviving wooden fragments from burials and on clay models

The home of the Asiatic nomad was either the ox-drawn wagon or the tent, both of which were commonly roofed with felt. The design of neither appears to have changed greatly over the millennia, apart from the use of spoked wheels.

a development carried out on the steppe, in which case the chariot, too, was in all probability first evolved in this area. At a still later date, the use of the horse for riding, the gradual development of the saddle and finally of stirrups were all unquestionably innovations for which the Nomads were totally responsible.

Again, as we have already seen, the composite bow must be regarded as an invention that went hand in fist with the growth of chariots and cavalry, and there can be little doubt that the higher civilizations of the Near East and of China both acquired this weapon from the same source. Since it would be difficult to build chariots of any quality or to make sophisticated weapons such as the composite bow without metal tools it must be assumed that the Nomads became as advanced in their metal-working techniques as their contemporaries in Mesopotamia, while there can be little reasonable doubt that it was through their contacts with the Nomads that the Chinese acquired a knowledge of bronze-working even if they themselves applied it to the making of ritual vessels rather than to the manufacture of tools and weapons. As we have already seen, early metal technology was in no way incompatible with the Nomadic way of life, and there is nothing outrageous in supposing that within each Nomadic society was to be found a number of bronzesmiths. Indeed, our limited knowledge of bronze tools from this vast area shows that this was the case.

What is less certain, however, was the role played by the Nomads in the transmitting of iron technology, or whether indeed they may not themselves have been responsible for the ultimate developments that made it possible to use iron on a large scale. The important factor in the evolution of iron technology was, of course, the introduction of the bellows, and it is of interest to notice that this device in its improved form appeared both in China and in the Near East at much the same time. One is bound to ask, therefore, whether the Nomads had themselves already perfected this instrument before it was known in either of the main areas of sedentary civilization. At the risk of being accused of romanticism, it may be pointed out that Anacharsis, who reputedly invented the bellows in Ionia, was himself a Scythian and would, had such a device already existed, have been well aware of it in his own homeland. Furthermore, of all the peoples of antiquity most likely to have developed this device, the Nomads with their long tradition of working with hides and leather, would seem to be the most probable candidates.

260 A silver appliqué ornament from Ordos, about 300 B.C.

Because of their way of life the Asiatic nomads were unable to practise those crafts requiring permanent work-shops. In other fields, such as metallurgy, they excelled, developing their own distinctive forms of weapons, tools and ornamentation, of which this seated mule may be seen as a typical example.

Unfortunately, this argument can immediately be countered by considering the development of the cross-bow, which again appeared at much the same time both in the Far East and the Near East, yet which in China was apparently developed principally to withstand the onslaught of the Nomads themselves. It seems highly unlikely that, in this case, the Nomads were the agents for transmitting the invention or that they were responsible in the first place for the evolution of the cross-bow itself.

Essentially the problem of assessing the contribution made by the Nomads to the development of technology in the ancient world lies in the fact that one can never be quite certain whether ideas were transmitted by the Nomads or whether new developments which were similar in character, but took place at great distances apart, were essentially independent inventions. Very much more will have to be known about the early history of the Nomads before a realistic appraisal of their contribution to technological advance during this period can be made.

The New World

It is the author's view that, correctly speaking, the development of technology in the New World, since in all probability it was totally unrelated to that of the Old World during the period under review, should have no place in this book. A very summary survey of a few aspects of technological evolution in the New World is included solely to illustrate the point just made. The difficulty of accepting the independent invention of a number of techniques may give rise to elaborate theories of past social contact to account for their appearance in widely dispersed areas. Briefly, it has been supposed by a number of writers that the early development of metal technology in the New World was the outcome of contacts made across the Pacific, and the purpose of this survey is to show that this was not necessarily the case.

The nub of much technological innovation in the New World appears to have been one of the least propitious areas of all – the coastal belt of Peru and Ecuador. This narrow strip of land, varying in breadth from twenty to forty miles, is largely desert, punctuated at intervals by rivers whose sources lie high in the Andes. Ultimately each river valley was to become virtually a separate state connected to its neighbours only by limited routes across the desert. Of the early development of this region we know very little indeed, but by 1000 B.C. the valleys were clearly occupied by settled agricultural communities. Like the people of the Near East, they were building both in stone and in mud-brick. They had cultivated the potato and maize as staple crops, but the wild fauna was less satisfactory for domestication than that of the Old World. Nevertheless, the llama, the vicuña and the alpaca were all being exploited, either as pack animals or to provide wool, and the guinea-pig had been domesticated as a source of food. Looms, and indeed weaving techniques in general, had become highly developed, and the people of this area were producing some of the most remarkable textiles of antiquity.

Within the river valleys it was possible to grow crops by means of irrigation, but higher up in the foothills of the Andes agriculture was only possible by terracing, and it is questionable whether this was practised at this early date. The Andes themselves, however, were a source of prolific mineral wealth, with deposits of gold, silver, copper and tin, and it was to this mountainous area that the people in the river valleys had to look for their metals.

By 800 B.C. the people of the Peruvian coastal belt had already begun to work in gold and as far as one can judge this metal was being acquired in exactly the same way as it was first won in the Near East, namely by the panning of placer deposits. During the subsequent five centuries there developed a school of craftsmanship devoted entirely to the hammering of gold in which the metal was beaten into shape and in which decoration was applied by the use of punches. This so-called Chavin gold-work is of interest chiefly because it relied to a degree upon another technology, the working of stone. The early Peruvians had already become adept at shaping hard rocks by polishing and other abrasive techniques, and very frequently gold objects were made by

261 A gold bracelet from Peru of about 500 B.C.

262

hammering the metal into moulds made of stone. The quality of craftsmanship was truly remarkable, but for five centuries there appears to have been no technological advance whatsoever, and it would appear that the metal was seldom heated, at least not sufficiently to make it melt. The end of the fourth century B.C. in Peru saw a political upheaval which as yet our limited knowledge makes it impossible to define, but in all probability the Chavin settlements were overrun by newcomers from some other area. Whatever the cause of change, however, the effect was dramatic, bringing with it enormous advances both in the making of pottery and in metal-working.

It would appear that the new technological developments were confined to a number of nuclei each in a distinct river valley or in a number of associated river valleys, but the differences between the products of one region and another appear to have been due largely to artistic preferences rather than to the use of different techniques. Ultimately the more northerly group of people, the Mochica, were to gain political dominance over the whole area.

Hitherto, pottery, which in itself may be seen as an independent invention of the New World, was made by the simple techniques that one would associate with primitive societies anywhere in the world. At first it had been made by modelling by hand, and later by placing a ball of clay on a saucer that could be made to rotate fairly rapidly. Vessels were produced in this way that were not dissimilar from those made on a turn-table in the Old World. With the appearance of the Mochica and other contemporary peoples on the Peruvian coast, however, the better-quality pottery was made almost entirely by a sophisticated casting technique. The component parts of each vessel were shaped

262 Peruvian metal-founding, part of a drawing from a sixteenth-century codex

The earliest metal-work in the New World was shaped solely by hammering. Gold ornaments, such as this bracelet, first appeared in Peru about 500 B.C.

From about 300 B.C. onwards metals – chiefly gold, silver and copper – were being melted, cast, soldered and extracted from their ores. Curiously, Peruvian craftsmen found the same solution to raising the temperature of their furnaces as did metallurgists of the Near East more than two millennia previously (see Figure 46). The Peruvians were still using blow-pipes at the time of the Spanish Conquest. Tapping the molten metal directly from the furnace to the mould, as shown in this picture, was not universally practised, for crucibles of this and earlier periods are known.

263 A Peruvian gold beaker of about A.D. 1000

264 Diagram showing the stages by which a vessel of this type was made, based largely on detailed metallographic studies

As in the Near East, despite improved casting and soldering techniques, a large part of Peruvian metal output continued to depend upon hammer-work, thus standing in strong contrast to contemporary metal-shaping methods in China, from which quarter some authors see the inspiration for early Peruvian metallurgy.

263

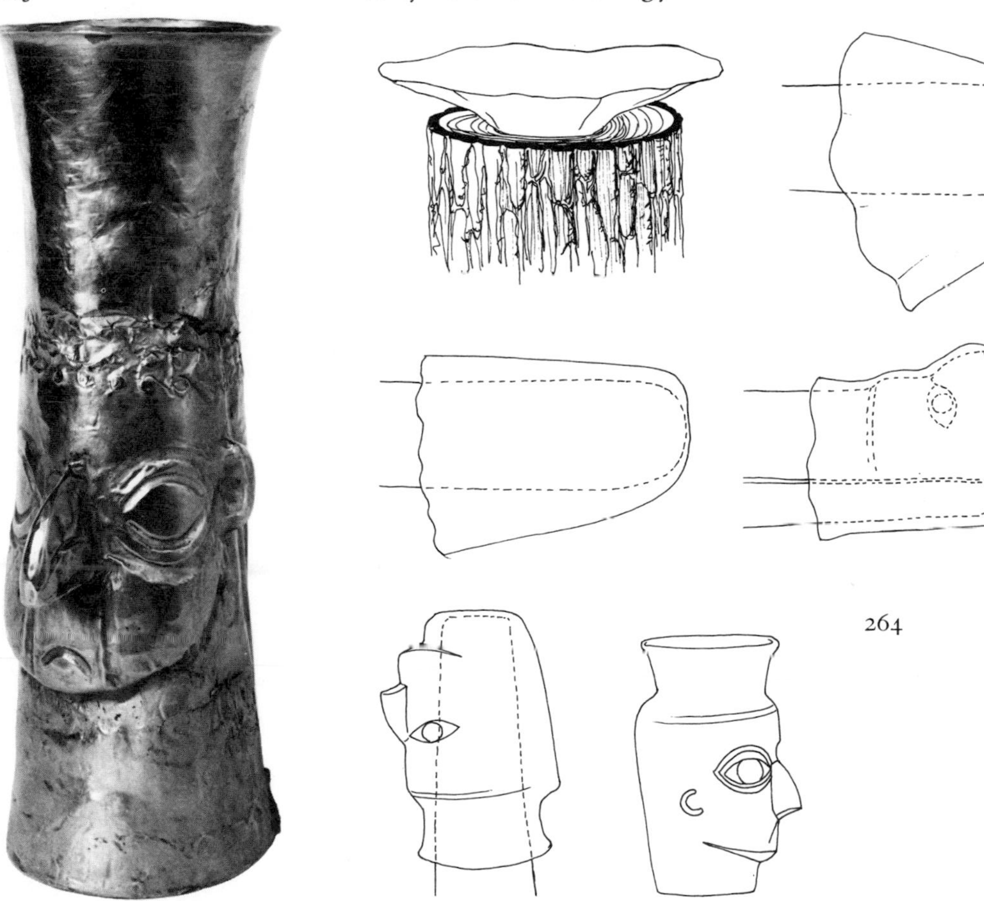

264

265 Diagram showing how Peruvian pottery was made in a mould from about 300 B.C. onwards

Pottery of high quality began to be made in Peru just before the beginning of our era. Slabs of clay were squeezed into opposing mould-pieces which were then joined to form the complete vessel. The technique, again, stands in total contrast to contemporary practice in China, and was almost without question a local development.

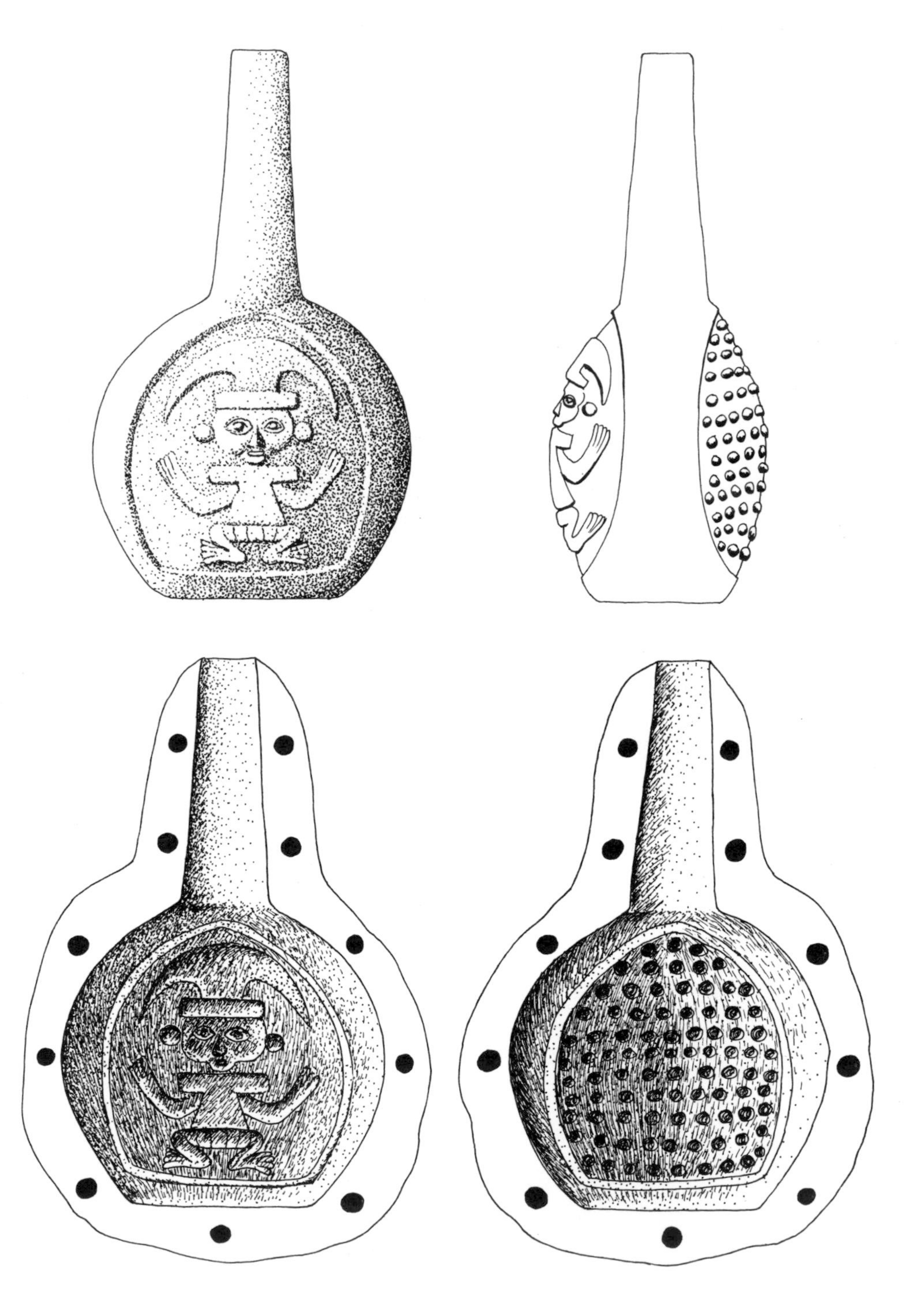

by forcing slabs of clay into open moulds, the various pieces being finally luted together to form the whole pot. Apart from the elaborately decorated wares themselves even the handles and spouts were made by this method. Sadly, however, although the Mochica pottery is a prolific source of information about their way of life no single vessel appears to give any hint as to how they set about working metals. As we have already noted in the Old World, however, stone moulds in which gold could be cast and later ceramic moulds were partly to pave the way to new metallurgical techniques.

Within the last few centuries B.C. in Peru were to appear in rapid succession the melting of gold, its casting, its alloying with silver and later with copper, the development of hard solder, the use of hammer-welding, and ultimately casting by means of the lost-wax method. During much the same period in Ecuador platinum and gold were being alloyed by heating together finely divided particles of the two metals – a technique that was later to be forgotten and not repeated until the nineteenth century. It was from this area that metal-working techniques were ultimately to spread to other parts of the New World. The diffusion, however, was slow and no metal-working, for example, was known in Mexico until the tenth century A.D.

To account for the rapid development of metal-working amongst the Mochica and in Ecuador some writers have suggested contacts across the Pacific from various quarters, as for example Japan, China and Indo-China. The view that such contacts existed is supported by a detailed examination of decorative motifs employed in either region, and indeed if one looks only at the decorative motifs the argument seems highly convincing. Broadly speaking the hypothesis is that in the period immediately prior to 300 B.C. a number of Chinese merchants made contact with the Peruvian coast in search of gold. After 300 B.C., due to political events in China, this supposed trade drifted to Southern China and Indo-China. During much the same period, theoretically, ships from Japan were trading with Ecuador.

Curiously, however, the techniques being used in the New World at this period were so utterly different from those current in the Old World that one must have serious reservations about the possibility of such contact. Thus, for example, while Chinese ceramics of this period were invariably thrown on a wheel those in Peru were being made, as we have just seen, by elaborate casting techniques. And while in China metal-work was being produced exclusively by casting, in Peru it was still largely being shaped by hammering. One might suppose that were the contacts sufficiently strong to allow the people of

Peru to adopt Chinese and Indo-Chinese decorative motifs they might, too, have acquired something of Chinese technology. Briefly, the answer to the apparently rapid development of metal technology amongst the Mochica probably lies far more closely at hand. Were one to know more about the people of the highlands of Peru and their development during these centuries as well as the nature of the political turmoil that ended so abruptly the earlier Chavin settlements the problem would in all probability not be one of serious debate. The Chavin period was one in which there was apparently great social stability but little technological advance. There followed a period of political chaos out of which, in a matter of a few centuries, grew a new society bursting with technological innovations. The brief survey of events in the Old World which follows will show a social situation with which we should by now be perfectly familiar.

Retrospect

We are now in a position to speculate upon some of the causes that led to technological evolution in the ancient world. The physical requirements were clear and obvious. An abundance of suitable raw materials and adequate communications were vital essentials. Areas lacking one or the other were not likely to be centres of technological advance. One might even postulate that the greater the variety of raw materials and the greater the ease of transportation so much the greater would in theory have been the rate of technological development. But the whole story was clearly not as simple as that.

Nowhere was the rate of technological advance a steady, even, upward climb. Always one seems to be confronted with sudden bursts of technological innovation followed by long periods of virtual stagnation, to be succeeded by yet another burst followed again by a long dormant period. In many places society has climbed to a particular level of technological advance and stagnated in this position even to this day. The neolithic societies of modern New Guinea and the Amazon are classic examples of this lack of advance, a failure which can be attributed entirely to a lack of suitable raw materials and to poor communications. But there were many other societies which, having climbed to a far higher technological level, then stagnated despite an abundance of raw materials and even excellent communications.

In effect, one of the limitations to technological evolution in antiquity seems to have been imposed by man himself, for certain social conditions were apparently inimical to further technological innovation. Indeed authoritarian governments aiming at stable social conditions appear to have been those under which there was least technological advance. The precise reason for such a state of affairs may have varied from one society to another. It may have been that the civil servants exercised far too rigid a control or did so in an unintelligent manner, while in modern terms there may have been too much capital investment in certain technologies to allow room for drastic innovations. Whatever the cause, however, the effect was the same, being tantamount to denying the benefits of good communications and resulting in a maintenance of the *status quo*.

The period of sudden technological evolution were to be seen in their most dramatic form at a time shortly after one of the stable societies had suffered a set-back at the hands of one of its less technologically advanced neighbours, and provided the kernel of the older society was left more or less intact it was invariably the intrusive population which showed the way to new technologies. Because they were aware of their lower technological standing they were willing to learn and because, being intrusive people, they had to be politically more adaptable, it was they and not the more stable civilizations that made the progress, for they were able to stand aside and look at what they saw with a critical eye and to select from their neighbours' technologies only what they wanted. Furthermore, such people had brought with them technologies of their own and by cross-fertilization, by applying techniques from one to another, they were able to introduce new concepts.

If the history of technology in the ancient world is to teach us anything at all it is to tell us that no technology can stand long on its own in glorious isolation. For a technology to thrive and to develop required the presence of other often quite unrelated technologies alongside it from which ideas could be lent and borrowed. Such was the case, anyway, in the ancient world. One is left wondering whether the world of today is so very different.

Note on Bibliography

The two main sources on which I have drawn during the writing of this book have been *A History of Technology*, Vols. I and II, edited by Charles Singer, E. J. Holmyard, and A. R. Hall (Oxford, 1954 and 1956) and *Studies in Ancient Technology* by R. J. Forbes (Leiden, 1955 onwards), of which so far nine volumes have been published. Between them these works list all the major references to the various topics up to the dates of their publication.

On the subject of the geographical background to the development of early technology the reader with probably find *The Geography behind History* by Gordon East (London, 1939) the most useful introduction; while the problems posed by Greek and Roman technology are very fully discussed in *Greek Science* by Benjamin Farrington (London, 1961).

In the period since *A History of Technology* and *Studies in Technology* first appeared a number of books have been published that add to and in some cases must make us modify our views on many subjects. Amongst these the *Dictionnaire archéologique des techniques*, Editions de l'Accueil (Paris, 1963) is useful since it covers China and the New World, areas barely touched upon in the former volumes.

The mechanics of mankind's evolution from hunter to settled agriculturalist has been brought under closer scrutiny than ever before in *The Domestication and Exploitation of Plants and Animals* edited by P. J. Ucko and G. W. Dimbleby (London, 1969); while the social changes that went hand-in-fist with developing technology are considered at length in *The Primitive World and its Transformations* by Robert Redfield (London, 1968). The machinery of change, both social and technological, is discussed fully in *Innovation: the Basis of Cultural Change* by H. G. Barnett (London and New York, 1953).

On a more limited front, *Chemistry and Chemical Technology in Ancient Mesopotamia* by M. Levey (Amsterdam, 1954) gives a tantalizing taste of the storehouse of information still locked up in the as yet unread Mesopotamian cuneiform texts.

The subject of early shipping in the Mediterranean is surveyed in its entirety in *The Ancient Mariners* by L. Casson (London, 1959) and in *Marines antiques de la Méditerranée* by Jean Meirat (Paris, 1964); while *Oared Fighting Ships* by R. C. Anderson (London, 1962) and *Greek Oared Ships* by J. S. Morrison and R. T. Williams (Cambridge, 1968) spell out in detail all that is really known about Greek and Roman oared craft. At the same time our understanding of the nature of early sea trade in the Mediterranean has been very greatly amplified by the detailed report of an underwater excavation in 'Cape Gelidonya: a Bronze Age Shipwreck' by George F. Bass in *Transactions of the American Philosophical Society*, **57** (1967), part 8.

The evolution of early wheeled vehicles has been discussed fully in *Die Landfarhzeuge des Alten Mesopotamien* by A. Salonen (Bonn, 1951), while fresh material from Central Asia is put forward in 'The Earliest Wheeled Vehicles and the Caucasian Evidence' by Stuart Piggott in the *Proceedings of the Prehistoric Society*, **34** (1968), pp. 266–318.

Many studies dealing with metal-working and ceramics in antiquity have appeared in the last decade. Of these *A History of Metals* by L. Aitcheson (London, 1960) gives, perhaps, the clearest account to date of the development of early metallurgy. *The Techniques of the Attic Potter* by J. V. Noble (London and New York, 1966) sets out in detail the changes that took place in the middle of the first millennium B.C. that were to dominate completely Greek and Roman ceramic technology.

The rapid development of iron technology in China during the critical centuries following the middle of the first millennium B.C. is fully discussed in *The Development of Iron and Steel Technology in China* by J. Needham (London, 1958); while all the major aspects of early Chinese technology are dealt with in *Archaeology in China* (four volumes) by Chêng Tê-K'un (Cambridge, 1959 onwards).

The development of early metallurgy in the New World is the subject of *Sweat of the Sun and Tears of the Moon* by André Emmerich (Washington, 1965). A shorter, but very useful account of the same subject-matter is to be found in 'Early Metallurgy in the New World' by D. T. Easby in *Scientific American* (April, 1966).

List of Illustrations

Index